CLIMATE CHANGE

Selected Titles in ABC-CLIO's
CONTEMPORARY
WORLD ISSUES
Series

For a complete list of titles in this series, please visit
www.abc-clio.com.

Books in the Contemporary World Issues series address vital issues in today's society, such as genetic engineering, pollution, and biodiversity. Written by professional writers, scholars, and nonacademic experts, these books are authoritative, clearly written, up-to-date, and objective. They provide a good starting point for research by high school and college students, scholars, and general readers as well as by legislators, businesspeople, activists, and others.

Each book, carefully organized and easy to use, contains an overview of the subject, a detailed chronology, biographical sketches, facts and data and/or documents and other primary-source material, a directory of organizations and agencies, annotated lists of print and nonprint resources, and an index.

Readers of books in the Contemporary World Issues series will find the information they need to have a better understanding of the social, political, environmental, and economic issues facing the world today.

CLIMATE CHANGE

A Reference Handbook

David L. Downie, Kate Brash,
and Catherine Vaughan

**CONTEMPORARY
WORLD ISSUES**

A B C CLIO

Santa Barbara, California
Denver, Colorado
Oxford, England

Library of Congress Cataloging-in-Publication Data
Downie, David Leonard.
    Climate change : a reference handbook / David L. Downie, Kate Brash, Catherine Vaughan.
        p.   cm. — (ABC-CLIO's contemporary world issues series)
    Includes bibliographical references and index.
    ISBN 978-1-59884-152-7 (hard copy : alk. paper)—
ISBN 978-1-59884-153-4 (ebook)
    1. Climatic changes—Handbooks, manuals, etc.   I. Brash, Kate.
II. Vaughan, Catherine.   III. Title.
QC981.8.C5D693   2009
363.738′74—dc22

                                                    2008042196

13   12   11   10   09   1   2   3   4   5

ABC-CLIO, Inc.
130 Cremona Drive, P.O. Box 1911
Santa Barbara, California 93116-1911

This book is also available on the World Wide Web as an eBook.
Visit abc-clio.com for details.

This book is printed on acid-free paper. ∞
Manufactured in the United States of America

Contents

List of Tables

List of Figures

Preface

Global climate change has begun. Average global temperatures have risen nearly 1°C since 1850, and in some places, changes are much more pronounced. Arctic temperatures, for instance, have increased 3 to 4°C over the past 50 years. Globally, 11 of the past 12 years are among the hottest ever recorded. These changes are impacting a range of human and natural systems. Scientists have observed changes in the timing of seasons; the range of plant and animal species; and regional patterns of precipitation, flooding, and drought. Sea levels are rising, and glaciers and Arctic sea ice are forging a steady retreat.

The Intergovernmental Panel on Climate Change (IPCC), an authoritative international body composed of thousands of scientists from 130 countries, has concluded that this warming is primarily the result of human activities. Since the time of the Industrial Revolution, activities including deforestation and the burning of fossil fuels have released increasing quantities of greenhouse gases into our atmosphere. These gases, which include carbon dioxide and methane, among others, trap heat that would otherwise escape into space. As such gases have accumulated in the Earth's atmosphere, they have intensified the natural greenhouse effect and are now causing climate change.

Looking toward the future, the IPCC reports that temperatures will rise 1.1 to 6.4°C by the end of the 21st century, with the range largely dependent on future greenhouse gas emissions. The type and severity of impacts that are associated with such temperature increases will vary by region, but on the whole they are expected to be negative—and, in some cases, disastrous. Furthermore, the greater the temperature increase, the greater the impacts we can expect. Fragile ecosystems, coastal areas, and low-lying islands will be destroyed. Species unable to adapt to changing conditions will go extinct. Agricultural pests and vector-borne diseases will spread, and people will suffer as droughts, floods, and storms become both more frequent and more intense. The world's poor will be hit first, and hardest, as changing climatic conditions exacerbate problems of food security, water scarcity, and sanitation.

What to do? Answering this critical question requires a basic understanding of climate science and policy, significant historical developments, current controversies and debates, proposed solutions, important people and organizations, and publications and Web sites to find additional information on this rapidly evolving subject. Part of the ABC-CLIO Contemporary World Issues series, this book addresses these topics following a specific structure designed to provide an accessible starting point to comprehending their many complex issues.

Chapter 1 provides the background and history necessary to understand the Earth's climate system, how it is changing, why human actions are largely responsible, what impacts will occur, and what policies have been adopted by the United States and internationally to address climate change. Chapter 2 explores important problems, controversies, and solutions. We focus on how debates about scientific, economic, and policy issues affect developments in the United States; we also address a suite of potential policy options for addressing climate change. Chapter 3 considers international issues, exploring how other countries approach this inherently global issue, including what impacts they face from climate change and the key components of their domestic and international policies.

Chapter 4 provides a chronology of key developments in the science, politics, and economics of climate change. Chapter 5 offers biographies of 28 particularly important people in climate change research and policy, many of whom are alive today. Chapter 6 contains essential data and documents for understanding the climate issue. The first section in this chapter includes graphs, tables, figures, and maps that illuminate the nature of the climate problem and the expected changes, including some that have become iconic symbols within the issue. The second section includes excerpts from key documents related to particularly important parts of the climate issue, such as IPCC reports, global climate treaties, and U.S. policy developments.

Chapter 7 provides information about some of the most important organizations involved in climate science and policy, including treaty secretariats, intergovernmental organizations, governmental agencies, research centers, environmental groups, and business organizations. Chapter 8 lists other resources—books, articles, reports, Web sites, and videos the reader can consult to find additional information on specific topics. Finally, the Glossary defines key terms and acronyms.

This book would not have been possible without the assistance of many people. Most important, we thank Kim Kennedy White and the entire team at ABC-CLIO for their patience and encouragement throughout the development of this volume. We would also like to thank Daniel Bader, Yevgeniya Bukshpun, Gang He, Linda Liu, Scott MacKenzie, Tatsiana Shakhmuts, Sofía Treviño, and Lyndon Valicenti for research and editorial assistance. We also are indebted to the Earth Institute at Columbia University and the many colleagues who supported our work.

Finally, we could not have completed this project, or any other, without the support and kindness of many family members and friends. We could never list all of those on whom we depend, but we are grateful for your kindness, intelligence, humor, friendship, and support. Kate Brash would especially like to thank Benet O'Reilly for looking on the bright side and for reminding her of what is most important. David L. Downie thanks his son, William, and daughter, Lindsey, for letting him use the computer and wife, Laura, for making his life better and better and generally putting up with him. Cathy Vaughan thanks Dan Green for keeping her head above water and for always seeing the forest for the trees.

1

Background and History

What is climate change? What causes it, and what impacts will it bring? What measures are being taken to slow it? What else should be done? These are obviously very important questions. This chapter helps to answer them by providing the reader with a basic understanding of climate science and policy. The chapter begins by outlining how the global climate system and the greenhouse effect work, how human activities are changing the climate, and what this change will mean for current and future generations. Then, the chapter examines measures being taken in the United States and around the world to address climate change.

The Global Climate System: An Introduction

The term *climate* refers to the average weather pattern of a planet or region over time. Its key components are temperature and precipitation. Though the climate of any particular location is affected by latitude, terrain, surface cover, and ocean temperatures, the climate system as a whole is driven primarily by gradients in solar radiation. Specifically, the curvature of the Earth causes the equator to receive more direct sunlight than the poles. The resulting temperature gradient creates a pole-ward transfer of energy, via wind and ocean currents, and shapes the global climate (Kump, Kasting, and Crane 2003).

The mechanism works like this: Intense sunlight warms the air above the equator, making the air molecules move faster and causing the air mass to expand. As a result, the air over the equator becomes less dense and exerts less pressure on the air columns that surround it. Because air moves from areas of high to low pressure, the pressure gradient associated with the differential heating of the Earth leads to horizontal movements of air. This air movement, called atmospheric circulation, carries heat energy from the tropics to the poles and is a primary component of global climate (Kump, Kasting, and Crane 2003).

Atmospheric circulation also helps drive circulation of the oceans. Together the circulation of the atmosphere and the oceans help determine temperature, precipitation patterns, and wind variability, which make up what we experience as the weather. Changes in average temperature alter these basic cycles, causing weather patterns to change (NASA EO 2008).

Global climate is also determined by the greenhouse effect. The greenhouse effect is the mechanism by which heat energy from the sun is trapped by the Earth's gaseous atmosphere. Short-wave radiation from the sun easily penetrates the Earth's atmosphere, warming the planet. Long-wave radiation emitted by the warmed Earth, however, is unable to pass back through the atmosphere. This trapped radiation warms the Earth's climate, making it roughly 33°C warmer than it would be without the atmosphere. An illustration of this process can be found in Figure 6.1 in Chapter 6.

The presence of certain trace gases in the Earth's atmosphere produces the greenhouse effect. Molecules of these "greenhouse gases" gain kinetic energy when they absorb long-wave radiation. This extra kinetic energy is then transmitted to other molecules, causing a general heating of the air.

Water vapor (H_2O) and carbon dioxide (CO_2) are the most abundant greenhouse gases. Methane (CH_4), nitrous oxide (N_2O), and ozone (O_3) are the other major naturally occurring greenhouse gases. Greenhouse gases exist in very small quantities in the atmosphere. Nitrogen and oxygen make up approximately 99 percent of the Earth's atmosphere but are not heat-trapping greenhouse gases. A complete list of greenhouse gases, including some human-made industrial gases, is found in Table 6.1 in Chapter 6.

The Carbon Cycle

Elemental carbon is a fundamental building block of life on Earth. It is an essential element in all living things and is spread in vast quantities throughout the Earth's system. The Earth's system maintains four major reservoirs of carbon: the geosphere, hydrosphere, biosphere, and atmosphere. The term *carbon cycle* is given to the natural movement of carbon between these reservoirs. The global carbon budget is the balance of all these transfers together.

The largest reservoir of carbon is the geosphere, which includes the Earth's crust, mantle, core, sediments, and fossil fuels, such as oil and coal. Although not a natural part of the carbon cycle, the primary way carbon leaves the geosphere is when people extract and burn fossil fuels. Carbon is also transferred from the geosphere to the atmosphere through volcanic eruptions.

The ocean, or hydrosphere, also contains vast quantities of carbon. Indeed, the deep oceans contain roughly 40,000 gigatons of carbon (Gt C), mostly dissolved in the form of bicarbonate ions like salt. Another 1,000 Gt C is found in the upper ocean, where it is rapidly cycled back and forth with the atmosphere. In regions of oceanic upwelling, carbon is released to the atmosphere when CO_2-rich water moves to the surface and interacts with the atmosphere.

The biosphere describes the Earth's plants, animals, soils, and surface. It holds roughly 1,900 Gt C. Carbon plays an integral part in maintaining the structure and viability of all living things, and CO_2 plays an essential role in the biosphere's circle of life. The terrestrial carbon cycle starts when plants absorb CO_2 gas from the atmosphere. During photosynthesis, CO_2 and water react in the presence of solar radiation to produce carbohydrates that form the structure of the plant. Animals that eat plants use the carbon to build their own tissue. Gaseous CO_2 is then released to the atmosphere when animals exhale, starting the process over again (NASA EO 2008).

The atmosphere contains roughly 750 Gt C, mostly in the form of gaseous CO_2. As mentioned above, carbon naturally and easily cycles in and out of the atmosphere from the hydrosphere and the terrestrial biosphere. Carbon also enters the atmosphere from the burning of fossil fuels, which originate in the geosphere. Although CO_2 accounts for just 0.04 percent of the atmosphere, it plays a critical role in sustaining plant life, which then sustains animal life. CO_2 is also central to the greenhouse effect because, like other greenhouse gases, it absorbs long-wave radiation and re-emits it downward, warming the lower part of the atmosphere.

This cycling of carbon through different reservoirs allows the Earth to maintain a natural carbon balance. When concentrations of carbon dioxide are upset—for instance, after a volcanic eruption—the system is able to gradually return to its natural state. The resilience of the system ensures that the carbon necessary for plant and animal life does not destabilize the oceans or the atmosphere. In the case of global climate change, however, the anthropogenic introduction of vast quantities of carbon, primarily from the burning of fossil fuels, has disrupted the carbon cycle (Kump, Kasting, and Crane 2003).

Energy Balance

Just as the Earth maintains a carbon balance, it also maintains an energy balance. The Earth's energy balance describes the interactive relationship, or balance, between the amount of energy that enters the Earth's system from the sun and the amount that eventually radiates back to space. As indicated earlier, some of this energy is trapped, warming the atmosphere and making our planet habitable.

The balance between energy in and energy out is also called the Earth's radiation budget. In part, the radiation budget is determined by the characteristics of the Earth's surface. Light colors reflect sunlight while dark colors absorb it. Accordingly, roughly 30 percent of the solar energy that reaches the Earth's system is reflected back into space by clouds and light-colored areas of the Earth's surface, including deserts and areas of persistent ice and snow cover. The remaining 70 percent of the sun's energy is absorbed by the atmosphere, land, and oceans (Dessler and Parson 2006). A change in the proportion of light-colored areas on the Earth's surface can alter the amount of energy reflected toward space and thus the energy balance.

The energy balance can also be altered if a change occurs in the amount of radiation received from the sun. In this regard, a disequilibrium can be caused by changes in either the sun's energy output or, in particular, the Earth's orbit, including its eccentricity, axial tilt, and precession. For example, changes in the orbit of the Earth are now known to be responsible for the 100,000-year ice age cycles witnessed over the last few million years (Kump, Kasting, and Crane 2003).

The final influence on the Earth's energy balance is the presence of heat-absorbing greenhouse gases in the atmosphere.

Greenhouse gases trap energy that would otherwise have been reflected back toward space. While the same amount of energy enters the system and the Earth emits the same amount back, an increase in greenhouse gases traps more energy within the atmosphere, heating the planet. Figure 6.2 in Chapter 6 provides an illustration of the Earth's energy balance.

Human Influence on the Climate System

The global climate is a complex system of interacting parts. The energy balance, carbon cycle, and greenhouse effect help maintain the climate as we know it. This section explores how human activities have begun to impact these mechanisms and change the climate.

Fossil Fuels and the Climate System

Fossil fuels such as coal, oil, and natural gas form when living organisms die and their remains are heated and compressed over extremely long periods of time. The oil and natural gas we use today were formed when aquatic plants and zooplankton accumulated on the sea floor hundreds of millions of years ago. In the absence of oxygen, which would have caused it to decay, the accumulated organic material gradually became buried under layers of sediment and, eventually, broke down into compounds of hydrogen and carbon called hydrocarbons. Coal, also a hydrocarbon, is formed by a similar process that occurs when terrestrial plants and animals decompose in swampy areas.

Fossil fuels produce significant amounts of energy when burned, making them important and valuable. However, burning fossil fuels also produces CO_2, introducing carbon that would otherwise remain trapped in the geosphere into the carbon cycle. Carbon released from the burning of fossil fuels and other human activities now adds an estimated 3.2 billion metric tons of carbon to the atmosphere annually (U.S. DOE 2008). Because the carbon cycle is not able to absorb all of the excess carbon, atmospheric concentrations of CO_2 have gone from 280 parts per million (ppm) in preindustrial times to nearly 400 ppm today. Carbon dioxide levels in the atmosphere are at their highest in at least 650,000 years (IPCC WG I 2007).

Coal

Coal is the most carbon-intensive fossil fuel. Burning coal releases more CO_2 per unit of volume than burning oil or natural gas. Coal reserves are abundant, well distributed, relatively inexpensive to access, and easy to use, so it is not surprising that the world uses a lot of coal. Coal-fired power plants produce nearly 40 percent of the world's electricity. Burning coal generates the heat needed for most cement kilns, iron and steel production (some 66 percent of steel production worldwide comes from iron made in blast furnaces using coal), and other industrial activities. Coal can also be converted into liquid fuel. South Africa, for example, has been converting coal to liquid fuel since 1955.

Economically recoverable coal reserves exist in vast quantities around the world. More than 50 countries mine coal, and existing reserves could last 200 years at current rates of extraction and use. The United States contains the world's largest known reserves (267.6 billion tons). China also possesses vast reserves and is currently the world's largest coal producer, extracting nearly 2.5 billion tons a year. Other important coal-producing countries include Russia, Australia, India, Indonesia, and South Africa (World Energy Council 2007). China and the United States are the world's largest coal consumers, respectively consuming 1.3 and 1 billion tons annually. Other countries that rely heavily on coal include Germany, India, Russia, and South Africa (EIA 2008).

Oil

Proven oil reserves are estimated at about 387 billion tons, or 1,315 billion barrels (World Energy Council 2007). Oil reserves are spread far more unequally throughout the world than coal. The Middle East contains about 61 percent of the world's proven reserves, followed by Africa with 11 percent, South America and Europe (including Russia) with 8 percent each, and North America with just under 5 percent. Oil reserve predictions, however, can vary widely, as different organizations sometimes provide different estimates. Also, it is possible that total reserves are larger—or smaller—than those currently proven.

The United States is the world's largest oil consumer (estimated at almost 21 million barrels per day), followed by the European Union (more than 14 million barrels per day), China (nearly 7 million barrels per day and increasing rapidly), Japan (5 million barrels per day), and Russia (3 million barrels per day) (CIA *World Factbook* 2008). Crude oil currently accounts for about

36 percent of global energy consumption, and few commodities are more central to today's economy. Remarkably versatile, crude oil can be converted into a variety of energy-related products including gasoline, diesel fuel, kerosene, heating oil, and motor oil. Crude oil also serves as a raw material in the manufacture of many plastics, fertilizers, tars, and lubricants.

Most forecasts predict that the global demand for oil will rise for several decades. Factors that could impact these projections include widespread efforts to replace the burning of oil products as part of stronger global efforts to address climate change; significant efforts to replace the import and use of oil on the part of the United States, Europe, and other countries due to geopolitical concerns; and product substitution in the face of very high prices. Price impacts can vary, however, as rising oil prices also make it economically viable to go to greater lengths to find and extract oil from areas previously deemed inaccessible, thus increasing supply and moderating prices.

Natural Gas

Natural gas currently accounts for about 23 percent of the world energy mix, ranking behind oil and coal (World Energy Council 2007). Gas-fired power plants supply significant amounts of electricity. Natural gas is also widely used for cooking and heating homes in the United States and Europe, as well as in the manufacture of fabrics, glass, steel, plastics, paint, fertilizers, and hydrogen, which in turn is used in the chemical industry, oil refineries, and hydrogen-powered vehicles.

Proven global gas reserves currently total about 173 trillion cubic meters and have grown at an average annual rate of 3.4 percent since 1980, largely due to improvements in exploration. The largest reserves are in the Middle East (about 41 percent of proven reserves) and Eurasia (32 percent, predominantly in Russia). North America has roughly 4.5 percent of global reserves, Africa has 8 percent, Asia has 7 percent, Central and South America have 4 percent, and Western Europe has 3 percent (World Energy Council 2007).

The top consumers of natural gas are Russia (about 610 billion cubic meters [cu m] annually), the United States (604 billion), and the countries of the European Union (nearly 500 billion). Other significant consumers include Iran and Germany (both consume about 98 billion cu m); Canada and the United Kingdom (roughly 92 billion each); and Japan, Italy, and Ukraine (about 80 billion each) (CIA *World Factbook* 2008). The use of natural gas has expanded rapidly over the last few decades—thanks to a number of techno-logical and exploratory breakthroughs and efforts on the part of the industry to make it price competitive (World Energy Council 2007).

Burning fossil fuels has released billions of tons of carbon into the atmosphere. Over the last 200 years, these emissions have come primarily from countries with advanced industrialized economies, sometimes called developed countries. Since the late 18th century, the United States, Europe, Japan, and Australia have relied on fossil fuels for energy. Developing countries, meanwhile, emitted relatively small amounts of greenhouse gases during this period, as their level of fossil fuel–based activity was not on the scale of that in the industrialized countries.

In recent years, however, developing countries have begun to increase their fossil fuel use. Many developing countries, while still very poor, are experiencing rapid economic growth and industrialization, following the pattern set in the 19th and 20th centuries by the United States, Europe, and Japan. As this process unfolds, most developing countries are understandably focused on raising living standards for their large populations and encouraging overall economic growth. Their rapid economic growth, combined with very large populations, has produced dramatic increases in greenhouse gas emissions. China has overtaken the United States as the world's largest emitter. India and Brazil are moving quickly up the global list. A map showing greenhouse gas emissions by country is included in Chapter 6 (see Figure 6.5).

Per capita greenhouse gas emissions in these countries are still much lower than those in developed countries, but their national emissions now have global importance. CO_2 emitted today will contribute to climate change for nearly a century before it eventually cycles out of the atmosphere. This delay means that climate change will continue even if we were to stop CO_2 emissions today. It also means that the emissions of developing countries, cumulatively still far lower than the historical emissions of developed countries, have the potential to greatly exacerbate the problem in the future. For this reason, many experts believe long-term emissions-reduction policies are necessary in both developed and developing countries.

Deforestation and Land-Use Change

Changes to the Earth's surface also contribute to climate change. Experts estimate that deforestation accounts for roughly 20 percent of global CO_2 emissions annually (IPCC WGI 2007). These emissions occur when the burning or decomposition of trees releases previously trapped carbon. Moreover, because forests take up roughly 2.4 Gt C each year as part of the carbon cycle, their destruction leaves fewer trees available to absorb carbon from the atmosphere, reducing an important carbon "sink" (in the context of climate change, sink is the term given to any any process, activity, or mechanism that removes greenhouse gases from the atmosphere) (UNFCCC 2007b).

Globally, the deforestation rate is roughly 32 million acres per year (UN FAO 2007). The World Bank estimates that almost half of the world's forest cover has disappeared in the last 200 years (World Bank 2007a). Forests are harvested for their wood or simply cut and burned to make way for agriculture and livestock. While the economic and political factors contributing to deforestation are complex, in general, harvested wood and saleable crops offer more immediate economic returns to developing countries than intact forests, although these returns may not be sustainable (Lindsey 2007).

Agriculture, which often replaces forests, also contributes to climate change, accounting for 10 to 12 percent of annual anthropogenic greenhouse gas emissions. About half of these emissions are methane (CH_4), produced primarily as a by-product of the digestion process of corn-fed cattle, rice cultivation, and management of livestock waste. Methane is 25 times more potent a greenhouse gas than CO_2. The other half of agricultural emissions comes from a variety of sources including nitrous oxide (N_2O) released from soils during cultivation and from the use of nitrogen fertilizers, CO_2 from farm machinery, and carbon released as overfarmed soils become depleted of organic matter (IPCC WG III 2007). Though a variety of techniques exist that can reduce the amount of greenhouse gases produced by agriculture, many are more expensive than conventional practices and are thus not widely employed.

The Industrial Revolution and Population Growth

Land-use change and energy consumption are the two most important drivers of anthropogenic climate change. But while humans have engaged in these activities for thousands of years, anthropogenic climate change is a relatively new phenomenon.

The scale on which humankind can impact the natural system has increased dramatically as a result of the rapid technological and social developments that occurred in the 18th and 19th centuries during the Industrial Revolution. It also contributed to a dramatic rise in population.

This historic period was characterized by the shift from manual to mechanical labor and the consequential reorganization of society. Because many of the era's most important technological and economic developments used coal—including the steam engine, steam locomotives, steam ships, steel and iron mills, heating systems for large factories, and coal-generated electricity plants—the Industrial Revolution also ushered in an era of heavy dependence on inexpensive fossil fuels.

In 1800, after the beginning of the Industrial Revolution, about a billion people lived on Earth. Today world population stands at more than 6.5 billion, and midrange scenarios predict an increase to more than 9 billion by 2050 (UNDESA 2006). By providing a relatively inexpensive input to economic growth, fossil fuels helped make this population boom possible. The growing population in turn spurred more demand for energy, agriculture, and deforested land. In this way, a population between 1 and 6.5 billion caused the climate problem. As population grows, the world will likely need enough energy and arable land to meet the needs and aspirations of 9 billion people. Whether we can do this without overwhelming the climate system remains unclear. More information about population, emissions, and climate change can be found in Figures 6.9, 6.13, and 6.14 in Chapter 6.

Why Does Climate Change Matter?

Though we sometimes talk about climate change as a problem for the distant future, in fact it is already under way. Eleven of the past 12 years were among the 12 warmest since people began keeping regular records. Global average temperatures between 2001 and 2004 were 0.76°C higher than they were between 1850 and 1899. The rate of temperature increase during the last 50 years is twice as high as in the 20th century as a whole. If current trends continue, temperatures will rise 1.1 to 6.4°C by the end of the 21st century; where the temperature actually falls on this range will be largely dependent on future releases of greenhouse gases (IPCC WG I 2007).

Climate change will affect a vast number of human and natural systems. The type and severity of impacts will vary from region to region, but on the whole they are expected to be negative—and in some cases, potentially disastrous. Some fragile ecosystems, coastal areas, and low-lying islands will be destroyed. Certain species will not be able to adapt to the changing climate and will go extinct. Many of the world's poorest people will likely experience more hunger, less access to fresh water, more exposure to extreme weather, and more threats from a variety of diseases.

The following section outlines the impacts of current and predicted climate change on human and natural systems. We divide them into seven overlapping categories commonly used in the literature: ecosystems, oceans and sea level, the cryosphere, freshwater, weather patterns, human health and security, and food security. Information in this section reflects findings of the second working group of the internationally recognized Intergovernmental Panel on Climate Change (IPCC) (IPCC WG II 2007), although other sources are noted when they provide additional details. Figures 6.15 and 6.16 and Table 6.2 in Chapter 6 also summarize the most likely impacts of climate change.

Systems Impacted by Climate Change

Ecosystems

An ecosystem is made up of plants, animals, and nonliving things, which interact to make a functioning natural system. The Earth as a whole is one ecosystem, and so are the Amazon region, Yellowstone Park, and your backyard. Healthy ecosystems are critical to the survival of many species and provide humans with a variety of benefits.

Changes in climate can put pressure on the different components and ecological relationships that are integral to particular ecosystems. Rising temperatures have already begun to affect individual species and systems. Indeed, in different parts of the world, scientists have observed changes in the timing of seasons; the range of particular plants and animals; and the local abundance of algae, plankton, and fish, particularly in high-altitude and high-latitude water bodies.

For example, the flowering of many trees in southern England now occurs an average of 4.5 days earlier in the spring than in the early 1990s. A number of insect species in different parts of the world have expanded their ranges northward over the past 20 years, including 37 species of butterflies and dragonflies, which have moved northward an average of 84 kilometers (52 miles) since the early 1980s. In addition, plankton, the basis of the North Atlantic food chain, have moved northward by 10° latitude over the past 40 years (IPCC WG I 2007).

While these changes may seem insignificant on their own, they represent important indications of larger changes to come. Some plant and animal species will be unable to respond quickly enough to changes in temperature, rainfall, or other habitat conditions and will thus die off. Species that can move to new ecosystems may crowd out others. Climate change will decrease the food available for some species, increasing competition and vulnerability as populations shift. In short, climate change could act as a stunning catalyst for species extinction. Many scientists estimate that 20 to 30 percent of the world's plant and animal species will be at risk for extinction if temperatures increase by more than 1.5°C (IPCC WG III 2007).

Oceans

Oceans play a critical role in regulating climate. Because they take much longer to heat than does land, oceans can store vast amounts of energy. The top 3.5 meters (11.5 feet) of the world's oceans have the capacity to store as much heat as the entire atmosphere. As a result, climate scientists estimate that heat absorption by the oceans has reduced by half the amount of warming that would have otherwise occurred in response to increasing greenhouse gas concentrations (IPCC WG I 2007).

The average temperature of the Earth's oceans (to a depth of 700 meters) increased by 0.10°C between 1961 and 2003 (IPCC WG I 2007). This may seem like a small amount until one realizes the immense amount of water involved and thus the consequential amount of heat required to warm it even by so little. Sea surface temperatures, although far more variable, also appear to be warming, in some places by much larger amounts.

Although the capacity of the oceans to store heat is enormous, it depends on the mixing of water from the deep ocean with surface waters, which takes place over a time scale on the order of 1,000 years. Other climate change–related factors, such as increased freshening from melting ice sheets and glaciers, may inhibit mixing, further slowing heat absorption. Slower heat absorption means that while the oceans will continue to absorb heat, they will not do so rapidly enough to prevent further warming of the Earth's surface and atmosphere.

The oceans also absorb CO_2. Oceanographers estimate that between 1750 and 1994, the oceans took up about 45 percent of all emitted CO_2. That figure has likely dropped by about 5 to 10 percent from 1980 to 2005, though a great deal of uncertainty surrounds this estimate. One thing that is known for sure is that the added CO_2 content is changing the chemical composition of ocean waters, causing them to become more acidic (IPCC WG I 2007).

This change in chemistry associated with increased CO_2 reduces the availability of calcium carbonate, which forms the shells and skeletons of corals and other sea creatures. Because coral reefs provide habitat and nurseries to many marine species, their decline from increasing ocean acidity and temperatures will increase the vulnerability of those species, which in turn will affect their predators. Rising temperature and CO_2 levels will have other serious consequences for the marine food chain as well (Doney 2007). For example, as the upper ocean warms and becomes less saline, the upwelling that brings cold, nutrient-rich water to the surface from the deep ocean will likely decrease. This decrease will reduce the biological productivity of the oceans. Changing temperatures will also alter preferred habitat locations and migration routes.

Sea levels are also rising in response to global warming. So far, this trend has been caused by the expansion of ocean water that occurs as it warms, but melting glaciers and ice caps could raise sea levels significantly in the future. Rising sea levels have the potential to affect many ecosystems and millions of people worldwide. More than 200 million people live in coastal floodplains. As many as 100 million live below sea level, 80 percent of them in developing countries (Dasgupta et al. 2007).

In Bangladesh, for instance, at least 35 million people live in areas that will be inundated if the seas rise by only 1 meter. In India 290 million people live on the coast. A sea level rise of only 1 meter would displace at least 7 million Indians and modestly higher levels would inundate the homes of tens of millions. In Vietnam, a sea level rise of 5 meters would submerge land inhabited by 38 percent of the population and accounting for 35 percent of economic production. A number of small island states in the Pacific and Indian oceans are no more than 3 to 5 meters above sea level at their highest point, and many other islands, such as those in the Caribbean, will suffer increased flooding, erosion, and risk of storm surges. Many docking facilities, airports, and other important economic assets on island nations are located only a few feet above sea level and will be greatly and negatively impacted by climate change (IPCC WG 1 2007). In the United States, where 53 percent of the population lives in coastal zones, a half meter sea level rise is expected to cost between $20 billion and $150 billion in relocation, infrastructure, and other costs. A sea level rise of 10 meters would flood about 25 percent of the U.S. population, in particular those living near the the Gulf and East coasts (Neumann, Yohe, and Nicholls, 2000).

The Cryosphere

The cryosphere refers to the parts of the Earth's surface that are cold enough for water to exist in solid form. It includes seasonal phenomena such as snow, river, lake, and sea ice, as well as more permanent features like glaciers, ice caps, ice sheets, ice shelves, and frozen ground. As a result of climate change, every element of the cryosphere has declined in size over the past 20 to 30 years.

Climate change has significantly impacted sea ice, alarming those who study it closely. Declining significantly since 1979, Arctic sea ice was 23 percent below the lowest previously recorded level at the end of the melt season in 2007 (NSIDC 2008). If ice continues to retreat at its current rate of 9 percent per decade, the Arctic Ocean will be completely ice-free by 2060, even in the winter. Since most scientists suspect that this rate will accelerate as more ice melts, it is predicted that the ocean will be without ice by 2040, if not sooner (Schiermeier 2007).

Melting sea ice creates a self-reinforcing cycle of warming in the Arctic. Because it is light in color, ice reflects 90 percent of incoming solar energy. Ocean water, however, absorbs 90 percent of the energy that hits it. Thus, the conversion of sea ice to ocean water creates a feedback effect that allows more heat to enter the system. This effect is the primary reason that Arctic temperatures have risen by 3 to 4°C over the past 50 years, nearly twice the rate of that of the rest of the world (ACIA 2004).

Climate change is also significantly impacting ice formations on land. Glaciers and ice sheets hold nearly 75 percent of the Earth's freshwater and occupy 10 percent of the world's land mass. Their fate is of great importance, as they contain water equivalent in volume to more than 60 meters of sea level rise (IPCC WG I 2007). The size of an ice sheet is determined by the balance between falling snow, which adds to the mass of the ice, and the outflow of glacial ice, which reduces it. In recent years, outflow has exceeded snow accumulation for both the Antarctic and Greenland ice sheets; moreover, the rate at which outflow exceeds accumulation is accelerating.

Many well-studied, midlatitude glaciers in mountain ranges in North America (including in Alaska, Montana, Wyoming, and Canada), South America (Andes), Europe (Alps), and Asia are retreating at historic and accelerating rates. If current rates of decline continue, for example, the vast majority of ice formations in Glacier National Park in Montana could disappear within 30 years. The ice at the summit of Mount Kilimanjaro, in Africa, formed more than 11,000 years ago and is expected to disappear between 2015 and 2020 as a result of human-induced climate change, even after persisting through periods of notable natural climate variability, including a 300-year drought (IPCC WG II 2007).

Another part of the cryosphere affected by climate change is permafrost. Permafrost is soil that remains at or below freezing for two consecutive years. Almost a quarter of all land in the Northern Hemisphere is permafrost. Degradation of permafrost occurs when parts of the active layer—the top layer, which may thaw in summer and refreeze in autumn—fails to refreeze. Eventually, the lower layers begin to warm as well, causing ground subsidence (sinking) and damage to any infrastructure that may have been built on top of the permafrost, including roads, pipelines, and buildings. In the past 20 years, the temperature of the active layer in the Arctic has increased by 3°C (IPCC WG I 2007). Significant permafrost melting has already occurred in some coastal areas of Alaska, and this temperature trend makes some experts worry that generalized thaws could become common.

As indicated earlier, one reason for the dramatic changes in the cryosphere is that many of its components are at or near the Earth's poles, where temperature changes are most drastic. Indeed, 40 percent of the cryosphere is found in the Antarctic and Greenland ice sheets and another 22 percent is found in the permafrost regions of the Northern Hemisphere, primarily the Arctic (IPCC WG I 2007). Rising temperatures in these areas have caused melting of the Greenland ice sheet, thawing of permafrost, locally significant coastal erosion, and dramatic seasonal reductions in Arctic sea ice (ACIA 2004).

Changes to the cryosphere will have a number of other far-reaching consequences. Less ice and snow will diminish the reflectivity of the Earth's surface, thereby increasing heat absorption and accelerating global warming. Melting ice and snow in the upper latitudes could reduce ocean salinity and alter ocean circulation patterns. Melting permafrost may release large quantities of methane and carbon that have been frozen for hundreds of thousands of years. A reduced cryosphere would also change wind patterns as the poles warm, reducing the difference between temperatures at the poles and the equator.

Freshwater

Freshwater is essential for the survival of plants, animals, and people. But global freshwater resources are declining due to increased population and pollution, and climate change will largely exacerbate the problem. The mechanism by which climate change will affect freshwater availability is known as the hydrological cycle—the continuous circulation of water from the Earth's surface to the atmosphere and back again. The hydrological cycle is expected to intensify under global climate change because warmer temperatures will increase both evaporation and the capacity of the atmosphere to hold moisture. More moisture in the air ultimately means more precipitation.

However, precipitation increases will not be distributed evenly. In fact, as a result of atmospheric circulation, increased precipitation will likely occur in areas that already experience high amounts of rainfall. In the world's driest areas, where water shortages are most acute, climate change will actually decrease water availability. In addition, the higher temperatures associated with climate change will contribute to increased evaporation, exacerbating dry conditions. As a result, places that already face frequent droughts, such as Australia, northern and southern Africa, Central Asia, and the American Southwest, will likely experience further reduced precipitation, which will lead to more frequent water shortages.

In the last century, precipitation increased 0.5 to 2 percent per decade in the midlatitude and tropical areas while declining by 0.3 percent per decade in the subtropics (IPCC WG I 2007). This trend is a concern for many people in the subtropics who rely on rain-fed agriculture for their livelihood. More frequent drought in this region will harm crops, increasing food shortages and, potentially, the incidence of famine. Declining rainfall will also limit water supplies for other purposes, especially in areas without adequate reservoirs. Increased precipitation in already-wet areas, on the other hand, will raise the threat of erosion, landslides, and floods, and potentially upset long-standing natural cycles on which irrigation, reservoir, and hydropower systems were built.

California provides an example of the impacts that changing rainfall patterns can have even in the wealthiest parts of the world. Most of California's freshwater falls as snow in the Sierra Nevada. When the snow melts in the spring and early summer, it fills up reservoirs that provide water for agriculture, household use, industry, and other purposes. In recent years, as a consequence of climate change, snowpack has decreased and melted earlier. Although the reservoirs fill earlier, they fill at a lower level and experience increased evaporation over a longer period, meaning that less water is available in late summer when demand is highest.

More ominous are the impacts that climate change will have on the water supplies of many developing countries. In addition to findings by the IPCC, World Bank research confirms that global warming is drying up mountain lakes and wetlands and melting glaciers in the Andes, threatening water supplies to major South American cities such as La Paz, Bolivia; Bogotá, Colombia; and Quito, Ecuador (World Bank 2007b). In India, rainfall and drought patterns are expected to become more intense, and extensive glacial melting in the Himalayan mountains could severely reduce freshwater supplies by midcentury (Mall et al. 2006). Similar changes are expected to occur in China as well.

Weather Patterns

Global warming will affect weather patterns in a variety of ways. Warming may shift seasons; rainfall patterns will change; and storms, heat waves, droughts, and floods will increase in number and severity. Hurricanes may become stronger. Meanwhile, many dry areas will see less rain and increased drought, while many wet areas will see more rain, stronger storms, and increased flooding.

Large, globally important patterns will change as well. For instance, the tropical Walker Circulation—which lifts air over Indonesia and causes it to descend toward Peru in a giant loop—is expected to slow as the global water cycle responds to increasing temperatures. As a result, air may circulate less vigorously in the tropical Pacific, creating conditions similar to that of the El Niño Southern Oscillation (Vecchi and Soden 2007).

El Niño is the name given to the warming of the tropical Pacific surface ocean that occurs every three to seven years, causing a shift in global weather patterns. Through a series of mechanisms, El Niño events are linked to increased rainfall across the eastern Pacific Ocean, Southwest United States, Mexico, and East Africa. Such events also produce drier than normal conditions in the Amazon River Basin, Colombia, Central America, southern Africa, and parts of the Pacific Northwest.

Though climate models agree that rising global temperatures will cause average climatic conditions in the Pacific to shift toward a permanent El Niño–like state, they remain divided as to whether El Niño events themselves will become more frequent or intense (Yeh and Kirtman 2007). However, either outcome would increase the likelihood of floods and droughts throughout the world and take weather patterns further away from those to which we have become accustomed.

Climate change has already increased the incidence of other extreme weather events unrelated to El Niño. Since 1950, the number of heat waves has increased, the extent of regions affected by droughts has increased, precipitation over land has marginally decreased, and evaporation has increased (due to warmer conditions). Generally, the number of precipitation events that lead to flooding have increased, though this is not true everywhere. Tropical storm and hurricane frequencies vary considerably from year to year, but evidence suggests substantial increases in intensity and duration have occurred since the 1970s (IPCC WG I 2007).

Human Health and Security

The World Health Organization estimates that climate-related impacts, including those associated with climate change, already cause 150,000 deaths a year and that this number will rise as temperatures increase (WHO 2007). Climate change affects human health by exacerbating other factors, including extreme events, weather patterns, agricultural productivity, the availability of freshwater, the distribution of pests and vector-borne diseases, and, potentially, resource conflicts.

As mentioned earlier, some of the most immediately noticeable impacts of global climate change will be in the form of changes in weather patterns. Storms, heat waves, floods, and expanded areas favorable to certain pests can harm people directly and affect food production. The massive heat wave in the summer of 2003, for instance, contributed to the deaths of as many as 50,000 people in Europe (EEA 2004).

Vector-borne diseases are transmitted via insects or other agents, sometimes with other animals serving as intermediate hosts. One example is malaria, which is transmitted to humans via certain species of mosquitoes. Each year 350 to 500 million cases of malaria occur worldwide, and over one million people die, most of them young children in sub-Saharan Africa (Center for Disease Control and Prevention 2008). Climate change will increase the distribution of mosquitoes able to transmit malaria, spreading the disease to areas where it is not currently found, potentially affecting millions of people with no immunity. Climate change will also extend the habitat of insects and other vectors that spread dengue fever, filariasis, schistosomiasis, and other diseases.

Some scholars and military officials have also warned that climate change has the potential to contribute to civil war and international conflict. Increased water scarcity and reduced agricultural productivity may add stress to preexisting social or political divisions, making conflict more likely. In the Darfur region of Sudan, for instance, increasing drought and desertification produced severe food and water shortages for nomadic groups, which, many observers believe, added to existing tension between herders and pastoralists. This tension has contributed to an ongoing military crisis, with hundreds of thousands of people killed or displaced (Ban 2007).

Food Security

Food production and security will be affected by climate change. Climate change will impact agricultural productivity, the stability of food supplies, and local access to food. These outcomes will increase the number of people at risk of hunger and starvation worldwide.

The impacts of climate change on food production will vary regionally. Overall, rain-fed agriculture in marginal areas in semi-arid and subhumid regions will be most at risk (Diouf 2007). In the tropics, crop yield is likely to decline with even small temperature increases (Diouf 2007). Increased climate variability, including more extreme events, is also expected to decrease production in the tropics. Areas that experience decreased water availability will face further difficulties producing and processing food. Temperature increases in higher latitudes are expected to extend growing seasons and may help increase production, but this positive impact will likely drop off once temperatures rise beyond 3°C. Climate impacts will also affect the distribution of agricultural pests, including locusts. If such pests find their way to new areas, they may overwhelm unprepared agricultural systems.

Given these threats, many scientists predict that climate change will very likely increase the number of people who experience food insecurity. These effects could be mitigated, however, by socioeconomic growth. Indeed, even in the face of dramatic climate changes, economic development and a slowed rate of population growth could reduce the number of people at risk of hunger in the coming century (Tubiello et al. 2007).

A History of Climate Change Science and Politics

Though anthropogenic climate change is caused by activities people have engaged in since the Industrial Revolution, very little has been done to forestall its impacts. One reason for the absence of an effective response is that, although the basic elements of the problem were first discussed in the 19th century, climate change science has only recently become well understood. Another major reason that little has been done to combat climate change is because climate policy and related economic issues are complex and divisive. To provide an introduction to these interrelated topics, this section outlines the history of climate change science and the development of U.S. and global climate policy.

Climate Science Pioneers

In the 1820s, the French mathematician Joseph Fourier published an article that analyzed the role of the atmosphere in warming the surface of the Earth (1827). Fourier recognized that while the Earth must re-emit energy from the sun, some of that energy must be retained in order to account for the temperature at the Earth's surface. He compared the effect of the atmosphere to the warming of air in a glass-sided container, giving rise to the green-house analogy that persists to this day.

In 1862, the Irish scientist John Tyndall took these conclusions one step further. After inventing the spectrophotometer, an instrument that allowed him to measure the relative transparency of different gases to different kinds of light, Tyndall conducted experiments that proved that CO_2, water vapor, and several other gases absorbed heat radiation. He speculated that variation in their atmospheric concentrations could be the cause of "all the mutations of climate which the researches of geologists reveal" (Fleming 1998, 73).

Swedish scientist Svante Arrhenius followed up this work by conducting a series of calculations to determine if levels of CO_2 in the atmosphere could affect the balance between incoming and outgoing radiation. His results, published in 1896, demonstrated that variations in atmospheric CO_2 could affect the energy balance, alter temperatures, and thus impact climate. Although Arrhenius considered it highly unlikely that human beings could alter the atmosphere enough to cause substantial changes, his colleague Arvid Högbom subsequently confirmed that changes in atmospheric composition large enough to affect the climate were theoretically possible.

However, because Arrhenius's research seriously oversimplified the climate system, including failing to analyze the role that oceans play in regulating CO_2 levels, his peers found it easy to dismiss his conclusions regarding the global warming potential of CO_2. As a result, the idea was more or less abandoned for decades until the 1930s, when a British engineer and amateur meteorologist named Guy Stewart Callendar began investigating temperature trends (Weart 2004).

In his 1938 paper "The Artificial Production of Carbon Dioxide and Its Influence on Temperature," Callendar predicted that temperatures would increase by 0.03°C per decade as a result of human CO_2 production. Though his work left out many significant factors, it was very important in reintroducing the idea of greenhouse warming into the scientific community (Charlson 2007).

In the 1950s, Roger Revelle, an oceanographer at the Scripps Oceanographic Institute in California, took another critical step when he showed that the world's oceans could not absorb as much CO_2 as scientists had previously assumed. Concerned about the possibility of greenhouse warming, Revelle helped secure funding to establish CO_2 monitoring stations across the globe as part of the activities associated with the 1956 International Geochemical Year.

One of the best monitoring locations, due in part to its relative isolation, was a weather observatory run by Charles David Keeling atop Mount Mauna Loa in Hawaii (Weart 2004). After only two years of monitoring, Keeling's data showed increasing atmospheric CO_2 concentrations. The data confirmed a discernable human influence on the composition of the atmosphere and thus the climate system. Keeling's data were used by other scientists to study the carbon cycle and to create simple models to simulate the movement of carbon between reservoirs, in particular, the oceans, the atmosphere, and terrestrial plants. Today, the graph of the Mauna Loa data is an iconic image in the field of climate science known as the Keeling Curve. An example is provided in Figure 6.4 in Chapter 6.

In 1961, the U.S. Department of Meteorology released a report stating that though global temperatures had increased until about 1940, they had been in decline since. This confounded even the authors of the report. The report acknowledged that greenhouse gas concentrations were increasing, and that rising greenhouse gases should lead to increasing temperatures. They were unable to offer an explanation for the apparent contradiction (Weart 2004).

During this period, many scientific questions concerning key elements of the climate system remained. Experts could not agree on the ways in which the climate system behaved, nor on the precise impact of CO_2 on the system. Moreover, there was no way to explain the apparently contradictory temperature trends. There was also no credible information on the extent of past climate change. Indeed, one of the only things the scientific community seemed confident of was that the climate system was probably less stable than previously assumed (Weart 2004).

Scientific Tools Advance

A number of important scientific advances helped to resolve this uncertainty. These advances included improvements in worldwide record keeping, satellite technology that allowed for better monitoring of the sun and Earth, a dramatic increase in computing power, and two other critical developments that provided scientists with crucial information and perspective.

The first of these advances was the development of methods to use ice cores to learn about the Earth's climate in the distant past. Scientists learned how to extract ice cores from areas in which snow and ice have accumulated for thousands or even hundreds of thousands of years without melting. By analyzing ancient air bubbles trapped when the ice formed, scientists gathered critical information on past climates (Biello 2004). Though the idea is relatively simple, perfecting the procedure was a significant breakthrough that afforded scientists a heightened understanding of climatic change.

In 1980, ice cores taken from Greenland revealed that CO_2 concentrations were 50 percent lower during the last ice age (15,000 to 22,000 years ago) (Delmas, Ascenio, and Legrand 1980). Five years later, in 1985, a team drilling in Antarctica gathered data that showed that CO_2 and temperature varied almost exactly in the same pattern, providing strong evidence that CO_2 and climate were closely linked (Lorius et al. 1985). The ice core record revealed that atmospheric CO_2 sometimes rose after significant changes in temperature, so it could not be said that levels of CO_2 directly caused the ice age, but the overall pattern clearly indicated a relationship between CO_2 levels and temperatures. Ice core records now extend back more than 650,000 years and have helped advance our understanding of past climates, as have analogous studies of cores taken from below the ocean floor.

The second critical advance in the quest to understand global climate change was the development of sophisticated climate models that use quantitative methods to simulate the interactions of the atmosphere, oceans, plants, and land surface. Scientists test climate models by running them against past conditions to see how well they "predict" what is known to have already occurred. After they are satisfied with the reliability of the models, scientists then run them with present conditions to see what future climates may look like.

The development of accurate climate models was a long process. As early as the 1920s, scientists began to search for mathematical equations that described the basic physics of the atmosphere and that would allow them to make accurate weather forecasts. Because weather is determined by a variety of interacting physical phenomena and a healthy dose of chaos, this project was extremely difficult. Indeed, it was not until the 1950s and the development of powerful computers that credible numerical weather prediction became possible.

At first, the models were capable only of regional weather forecasting. Scientists continued to develop modeling capabilities, however, seeking to create a general circulation model (GCM) that would simulate the circulation of the entire atmosphere in three dimensions. In 1955, Norman Phillips of Massachusetts Institute of Technology developed the first GCM. With ever-greater computer power and the increasing availability of empirical data that could be used as inputs in the models, GCMs became more sophisticated. In 1975, a team of scientists published a model that predicted that a doubling of atmospheric CO_2 would lead to a 3.5°C increase in global average temperature.

Today, GCMs are considered reliable enough to provide information on the likely response of the global climate system to possible changes in several parameters, such as increased concentrations of a particular greenhouse gas or decreased sea ice cover in the Arctic. These models have had an indispensable role in establishing the relationship between anthropogenic greenhouse gases and temperature trends, and in demonstrating the likely impacts of climate change on different parts of the world (AIP 2007). These models, and the newly developed ice core methods, significantly bolstered scientific understanding of past climate, the operation of the climate system, and potential future climate change as a result of human activity.

A History of Climate Change Policy

As evidence mounted, more people began calling for action to address it. These calls now emanate from all parts of the globe, although only a small number of countries, mostly in Europe, have actually implemented meaningful climate policy. The United States, which pioneered modern environmental policy in the 1970s, has not been among the leaders in addressing climate change.

In general, policy to address climate change can take two broad forms. The first is mitigation—efforts to reduce the amount of climate change. Mitigation policies focus on reducing net greenhouse gas emissions by preventing or capturing emissions; using non-fossil-fuel energy sources such as wind, solar, and geothermal; increasing energy efficiency; altering agricultural practices; enhancing sinks by reducing deforestation and planting and protecting new forests; and many other methods.

The second strategy is adaptation—efforts to make the impacts of climate change less harmful to humans and ecosystems. While the list of potential adaptation strategies is endless, prominent adaptation policies include preparing coasts and river basins for rising sea levels and storms; combating the potential spread of particular pests and diseases; improving the management of lakes, reservoirs, and rivers to combat water shortages; helping farmers understand and prepare for changing weather conditions to protect agricultural productivity; planting drought-resistant trees and crops; and creating information systems to alert people in advance of particular impacts. For many years, mitigation options dominated climate policy discussions, but recently, adaptation has become recognized as an equally important topic. Most experts now believe that both strategies should be pursued simultaneously in order to reduce the climate change impacts experienced by current and future generations (Kane et al. 2003; IPCCWG II 2007).

Significant research exists on mitigation and adaptation, but the creation and implementation of strong emissions-reduction policies have been slow. The next section outlines the history of climate policy in the United States and the international treaty processes.

Climate Change Enters U.S. Politics

Climate change slowly entered the public arena as scientific information on the issue improved. In one of the most important initial developments, the publication of Keeling's data helped prompt the President's Science Advisory Committee to hold the first hearing on climate change in 1965. This hearing was followed by a month-long workshop on climate change at the Massachusetts Institute of Technology in 1970 and an international conference in 1971.

In 1974, scientists discovered that chlorofluorocarbons (CFCs)—chemicals widely used as coolants in refrigerators and air conditioning systems, as propellants in spray cans, and as manufacturing inputs and industrial solvents—could damage stratospheric ozone. Stratospheric ozone, often called the ozone layer, helps protect the Earth from harmful ultraviolet radiation and is a critical component of the Earth's system. When it was discovered that CFCs damaged the ozone, Congress responded by banning nonessential uses, including in aerosol sprays in 1977. Though at the time the ozone issue had no connection with climate change, it illustrated that humankind could impact the atmosphere in fundamental ways and that scientific findings about a future atmospheric risk could arouse the public and government officials enough to produce meaningful legislation.

In 1978, Congress passed the National Climate Act, which established the National Climate Program office and charged it with conducting research on greenhouse gases and potential climate change. The program came under attack from the Reagan administration in the early 1980s, but lobbying led by Sen. Al Gore and reports by the Environmental Protection Agency and the respected National Academy of Sciences, which warned that global warming could cause dramatic impacts, convinced the Reagan administration to continue its funding (Weart 2004).

By the late 1980s, several climate-related bills had been introduced in Congress. In 1987, Ronald Reagan, who served as U.S. president from 1981 to 1988, signed the Global Climate Protection Act, which required the administration to prepare a plan to stabilize U.S. greenhouse gases (Weart 2004). However, while reports were written, no substantial action came out of this initiative. And though Reagan's successor, George H. W. Bush (1989–1992), pledged to take action on climate change during a 1988 campaign speech, once elected he largely focused on the uncertainties surrounding the climate problem, possibly because of lobbying by industry allies. The major exception to this inaction was U.S. participation in the global negotiations that produced the 1992 United Nations Framework Convention on Climate Change (UNFCCC), which the United States subsequently signed and ratified.

After Bill Clinton (1993–2000) took office, it appeared that Washington might initiate more substantial action to address climate change. The vice president, Al Gore, persuaded Clinton to endorse the proposed United States Climate Change Action Plan, which would have committed the nation to reducing greenhouse gas emissions to 1990 levels by the year 2000, in line with the nonbinding goal in the UNFCCC concerning industrialized country climate policy. However, fierce opposition from industries that would have been negatively affected by the regulations and their congressional allies stalled the initiative, and Congress instead passed legislation that included relatively modest steps to improve energy efficiency (Weart 2004).

Congressional opposition also prevented U.S. ratification of the 1997 Kyoto Protocol, an adjunct to the UNFCCC that included binding emissions reductions for industrialized countries. As negotiation of the protocol was nearing completion, the

U.S. Senate unanimously passed a Sense of the Senate resolution (known in this case as the Byrd-Hagel resolution) stating its opinion that the United States should not join any climate agreement that did not include binding emissions reductions for large developing countries or that would "result in serious harm to the economy of the United States" (U.S. Congress 1997). Because ratification by the Senate is required for the United States to become a legal party to any international treaty, this resolution served notice that Washington would not join in implementing global climate policy.

Shortly after taking office, President George W. Bush officially announced that the United States would make no further efforts to participate in the Kyoto Protocol. This announcement reflected the significant lobbying efforts of the fossil fuel industry, continued (but not universal) opposition in the U.S. Senate, as well as Bush's apparent personal views. The Bush administration also renounced any intention to place binding domestic restrictions on U.S. greenhouse emissions.

Subsequent disclosures revealed that the industry lobbying effort included an intentional disinformation campaign designed to cast doubt on climate change science, and that certain government officials had participated in this effort (U.S. House of Representatives 2007). Partly as a result of efforts to spread false information, many Americans were confused by or disinterested in climate change in the first few years of the 21st century. In addition, Americans had never before confronted such a complex environmental problem, which made communicating the gravity of the issue extremely difficult.

Scientific developments marched on, however, and the seriousness of the situation became increasingly difficult to ignore. In addition, industrialized countries that were already implementing mitigation measures in response to the Kyoto Protocol, as well as many developing countries, increasingly called on the United States to take action. Environmental groups and an increasing number of local and state officials followed suit. The Bush administration responded to these developments with a number of nonregulatory actions: advocating energy efficiency, calling for broader voluntary initiatives, and developing the Climate Change Research Initiative. The administration argued that this research would help resolve important uncertainties so that the United States could decide on a course of action. Critics called the move a delaying tactic.

During much of this period, the Bush administration refused even to acknowledge that human activities were causing climate change, despite the fact that this position was enshrined in the 1992 UNFCCC treaty that had been signed by the president's father and ratified by Congress. This stance did not formally change until mid-2004, when a federal report, cleared by the White House, stated that anthropogenic greenhouse gases were contributing to the warming trend seen in recent decades (CCSP 2004). President Bush himself did not acknowledge the link between greenhouse gases and climate change until 2005 (Revkin 2007).

Meanwhile, congressional attitudes began to change. In October 2003, Republican Senator John McCain (Arizona) and then-Democratic Senator Joseph Lieberman (Connecticut) cosponsored a bill to reduce carbon emissions using a "cap-and-trade" system, an emission trading strategy that had previously worked to reduce sulfur dioxide pollution from U.S. coal-fired power plants. The bill met strong opposition from the Bush administration and was denounced by many senators, but in the end it lost by just 12 votes, signaling that U.S. climate politics were changing (Weart 2004).

Outside of Washington, even more significant policy developments were taking place. Responding to the absence of federal action on climate change, a large number of local, state, and regional governments initiated efforts on their own. Some of the boldest steps came from California.

Perhaps most important is the California Global Warming Solutions Act of 2006, which requires the reduction of greenhouse gas emissions in California to 1990 levels by 2020 and to 80 percent below 1990 levels by 2050. Though the infrastructure necessary to monitor, verify, and enforce these reductions is still being established, the targets themselves are in line with internationally accepted recommendations for avoiding dangerous climate change (IPCCWG III 2007). Also, because the law includes penalties for noncompliance, it is the first enforceable greenhouse gas reduction mechanism in the United States.

By itself, California is one of the 10 largest economies and one of the 10 largest greenhouse gas emitters in the world. California's size gives its environmental policies political and economic influence, often persuading other states to take similar action. Moreover, the state's willingness to enact such aggressive measures while under the leadership of a Republican governor indicated to other states that concerns regarding the economic costs of initiating climate policy might be overstated (for more information on key economic controversies, see Chapter 2).

From 2006 through 2008, several other states, including Connecticut, Hawaii, Maine, Minnesota, and New Jersey, put emissions-reduction requirements into law. Many more states developed action plans requiring utilities to produce a certain percentage of electricity from renewable resources. A number of states also adopted California's vehicle emissions standards, which require 30 percent emissions reductions by 2016.

As of October 2008, 37 states take part in five different regional climate initiatives that require them to set a regional emission target and develop a market-based emissions trading scheme. Though they have different goals and reduction terms, these five initiatives—the Regional Greenhouse Gas Initiative in the Northeast; the Southwest Climate Change Initiative; the Western Climate Initiative; the West Coast Governors' Global Warming Initiative; and the Midwestern Regional Greenhouse Gas Reduction Accord—cover a large majority of U.S. emissions, more than half the country's land area, and a vast segment of the population (Pew Center 2008). In addition to their collective impact on emissions, the state and regional initiatives play an important role in stimulating national economic, technological, and political advances in climate change mitigation.

As this book goes to press, a number of climate bills are under discussion in the U.S. Congress. They follow on the heels of the Fourth Assessment Report of the Intergovernmental Panel on Climate Change (see below), which many credit with removing nearly all reasonable arguments for delaying action on account of scientific uncertainty. Many members of the House of Representatives and Senate, along with dozens of governors and hundreds of mayors, have publicly stated their support for mandatory, nationwide emissions reductions. In addition, dozens of chief executive officers from major U.S. corporations, as well as many leading scientific, religious, and cultural groups, have issued a string of consensus statements announcing their concern for the seriousness of the climate issue and their support for new U.S. and global policies (Downie, Gellers, and Raven 2006; see further discussion of this topic at the beginning of Chapter 2).

While President Bush and a number of Republicans remain opposed to mandatory controls, senators McCain and Barack

Obama, the 2008 Republican and Democratic presidential candidates, respectively, both endorse reducing U.S. greenhouse gas emissions by 50 to 80 percent by the middle of this century. While no certainty exists, most observers believe that the United States will see some type of binding national greenhouse gas emission reduction program signed into law by the end of 2009 or 2010.

Clearly, climate politics have evolved considerably in the United States during the last decade. It remains to be seen, however, if domestic action will be sufficient and implemented quickly enough to mitigate the most threatening climate impacts.

Climate Change and International Relations

European countries have taken far greater measures to limit their greenhouse gas emissions than has the United States. The difference is important because climate change is an inherently global problem and strong policy differences between the United States and Europe make solving it difficult. Significant policy differences also exist between most industrialized countries and the largest developing countries. These governments as well as many others must come to a consensus in order for the world to implement effective collective mitigation efforts.

Climate as a Global Problem

The climate is an international public good. We are all free to enjoy it, but protecting it requires the mitigation of greenhouse gas emissions from many different sources around the world. Because no one can be excluded from the benefits of climate change mitigation—and, conversely, because no individual can reap all the benefits of his or her own mitigation efforts—we end up investing less in mitigation than desired. Overcoming this problem will require significant international cooperation.

In some ways, the dynamics of the problem are quite simple. Imagine a virtuous individual struggling to reduce her carbon footprint. This individual installs solar panels; uses highly energy-efficient compact fluorescent lightbulbs, appliances, and insulation; recycles and buys products made from recycled materials; eats locally grown food and less meat; and rides a bike to work and buys a high-mileage car for long trips. She feels good about these efforts until she realizes that other members of the community are not working as hard to combat climate change. Driving sport utility vehicles, gobbling hamburgers, and downing bottled water, these less-virtuous neighbors take recreational vehicle jaunts to distant parts of the country, oppose tax incentives for green energy, and support a proposed coal-fired power plant to lower electricity rates, which they prefer over energy conservation.

The virtuous individual is discouraged. Any positive impact that her efforts might have on the global climate are being negated by her neighbors' greenhouse-gas-intensive lifestyle. If she could somehow consolidate her positive impact—ensuring that she, at least, was protected from future climate change—then she would continue the mitigation efforts. As it is, she wonders why she is trying so hard when she will surely still suffer the consequences of climate change because of actions by others in the community.

Now consider the opposite scenario: a less-virtuous individual who refuses to mitigate even as those around him do. While this individual continues to pollute the global atmosphere, others in the community drive less or more efficiently, consume wisely, and make use of renewable energy. They vote for climate-friendly policies, support reducing emissions from their utilities, and stop using carbon-intensive consumer products. In this situation, our less-virtuous individual also has no incentive to reduce emissions. Regardless of his own behavior, he will still benefit from the climate-friendly conditions in which he lives.

In many ways, international climate interactions are more or less the same as those of the individuals described above. Some countries otherwise inclined to invest in mitigation measures are discouraged when other nations do not agree to make similar cuts. Meanwhile, some countries—aware that they will benefit from the efforts of others—decide to "free-ride" and refuse to take mitigation steps of their own. (In economics, the free rider problem refers to a situation where some individuals in a population either consume more than their fair share of a common resource, or pay less than their fair share of the cost of a common resource.) Even if all countries agree that climate change is a problem and desire a certain level of mitigation, addressing the problem effectively will be difficult unless they agree to work together and can structure an agreement in such a way that everyone has confidence in the commitments of others. For this reason, an international agreement—the United Nations Framework Convention on Climate Change—stands as the centerpiece of the global effort to combat climate change.

The United Nations Framework Convention on Climate Change

International environmental agreements have been around for a long time. For reasons similar to those described above, several countries agreed to work together to protect halibut reserves and their marine environment in the Pacific Ocean as early as 1923 (IPHC 2007). More recently, nations have created a series of international environmental agreements to address issues that no single nation could successfully tackle alone. These agreements include binding treaties on issues as diverse as protecting endangered species, managing hazardous wastes, reducing air pollution, restricting production and use of toxic chemicals, protecting the ozone layer, safeguarding biodiversity, and reducing pollution in the oceans and regional seas (Chasek, Downie, and Brown 2006).

Given this history, it is not surprising that as scientific knowledge regarding the causes and potential consequences of global climate change has increased, many observers and some governments began to call for consideration of coordinated international action. The nature of the climate problem made it clear that no single country could tackle the issue successfully and that global scientific and political cooperation would be necessary to understand it and create an effective response.

One of the most important early steps toward a coordinated effort took place in 1988 at the World Conference on the Changing Atmosphere. After discussion on a variety of environmental issues relating to the atmosphere, governments requested the United Nations Environment Programme (UNEP) and the World Meteorological Organization to create the Intergovernmental Panel on Climate Change (IPCC). The IPCC was not intended to be a research organization but rather to provide comprehensive assessments of the current state of scientific knowledge on climate change in order to inform international policy makers.

The IPCC issued its first report in 1990. This interim report (it was updated comprehensively in 1992) confirmed the natural greenhouse effect and asserted that greenhouse gas emissions from human activities could enhance its impact. It reported an observed increase in global temperatures of 0.3 to 0.6°C over the past 100 years and predicted a 1 to 3°C increase by 2100. The United Nations General Assembly took official note of these findings and, using the IPCC interim report as scientific justification, authorized formal negotiations on a global climate agreement. Negotiations began in 1991 and, informed by the suplemental

IPCC assessment released in 1992, quickly resulted in the UN-FCCC, signed by 154 nations at the Earth Summit in Rio de Janeiro, Brazil, that year.

The UNFCCC stands at the center of global climate policy. Its stated goal is to "prevent dangerous anthropogenic interference in the Earth's climate system" (UNFCCC 1992, Article 2). Reflecting industrialized nations' greater contribution of greenhouse gases to the atmosphere, the parties to the convention—both industrialized and developing nations—agreed in principle to recognize their "common but differentiated responsibilities" to achieve this goal (UNFCCC 1992, Article 4). To facilitate their doing so, the convention encouraged industrialized countries to pursue the voluntary nonbinding goal of reducing greenhouse gas emissions to 1990 levels by the year 2000.

The Kyoto Protocol

When they adopted the convention, governments knew that it would not be enough to produce the emissions reductions necessary to prevent significant climate change. At the first Conference of the Parties (COP 1), held in Berlin in 1995, parties launched a new round of talks to decide on stronger and more detailed commitments for industrialized countries. After two and a half years of intense negotiations, they reached agreement on the Kyoto Protocol on December 11, 1997, at COP 3 in Kyoto, Japan.

The United States' refusal to ratify the Kyoto Protocol was a serious blow. To enter into force, the treaty required the signature of 55 nations, representing 50 percent of the world's greenhouse gas emissions. With the United States accounting for more than 20 percent of global emissions, reaching this goal became much more difficult. Nevertheless, delegates continued to meet to hash out many of the details of Kyoto's implementation. In November 2001, they finalized these details in the Marrakech Accords, after which many nations agreed to sign on. In October 2004, the treaty's unique entry-into-force provision was finally met when Russia ratified the protocol. It came into force 90 days later, on February 16, 2005. To date, 181 countries and the European Community have ratified the agreement.

The Kyoto Protocol requires industrialized countries to reduce their collective emissions of the six most important greenhouse gases by at least 5.2 percent below their 1990 levels by 2008–2012. Within this group requirement, countries negotiated individual targets. Some countries were required to reduce emissions by more than 5.2 percent, while others were allowed to grow their emissions in a controlled way. For instance, Australia (which did not ratify the treaty until 2007) argued that because its economy was far more heavily dependent on fossil fuels than the average industrialized country, it should not have to reduce as deeply. Thus Australia was tasked with limiting the growth of fossil fuel emissions to 8 percent above 1990 levels. Meanwhile, the United States, which has still not ratified the treaty, accepted an emissions-reduction target of 7 percent below 1990 levels, while both Canada and Japan agreed to reduce emissions by 6 percent.

The European Union (EU) as a whole agreed to reduce its emissions by 8 percent. In order to achieve this reduction, the EU negotiated a burden-sharing agreement to allocate emissions reductions among its members. Germany, for instance, accepted a reduction target of 21 percent; the United Kingdom agreed to reduce by 12.5 percent; France committed to keeping emissions at 1990 levels; and Sweden committed to keeping emissions growth to 4 percent. Norway, which is not part of the EU, has a Kyoto target requiring emissions to increase no more than 1 percent above 1990 levels (Chasek, Downie, and Brown 2006).

In addition to this burden-sharing scheme, the Kyoto Protocol contains several flexibility mechanisms aimed at reducing the cost of emissions reductions and promoting environmentally sustainable economic development in poor countries. The details of these mechanisms were not finalized in Kyoto but were completed at subsequent COPs.

The first flexibility mechanism is an International Emissions Trading scheme, which created a market for countries to trade emissions with one another. Theoretically, this lowers the overall cost of emissions reductions by allowing those who can reduce emissions most cheaply to sell emissions credits to those for whom reductions are more costly. The second flexibility mechanism is Joint Implementation, which allows developed countries with commitments to cooperate on projects and transfer emissions allowances on the basis of such projects. A third is known as the Clean Development Mechanism (CDM), which allows developed-country parties to finance emissions reductions in developing countries and count some of those reductions toward their own commitments. Because the projects take place in countries with lower levels of development, where technologies are older and less efficient, the mechanism is intended to produce cheaper emissions reductions. It has the added stated advantage of helping poor countries achieve environmentally sustainable development paths via diffusion of clean energy and energy-efficient technologies (UNFCCC 2007a).

By creating mandatory targets, addressing six major greenhouse gases, allowing each country to decide how to reach its own target, giving credit for sink expansion, and creating the flexibility mechanisms, the Kyoto Protocol creates a comprehensive yet flexible greenhouse gas emissions-reduction regime. Although far from perfect, it sets an important precedent for international cooperation on climate change and speaks to the seriousness with which many of the world's governments view the climate issue (Aldy, Barrett, and Stavins 2003).

The Kyoto Protocol was never seen, however, as the final step. Kyoto does not mandate deep or long-term emissions reductions, nor include emissions from large developing countries; it has not attracted participation by the United States, and still has not developed sufficient incentives within its flexibility mechanisms to assist developing countries in taking stronger action to prevent deforestation. Thus Kyoto is only the beginning of a long-term international effort to address the climate problem.

The current commitment period of the Kyoto Protocol ends in 2012. No agreement has been reached on what will happen after 2012. During COP 13, held in Bali, Indonesia, in December 2007, governments did agree on a schedule designed to guide negotiations toward conclusion of a new comprehensive agreement in December 2009. This "road map" included written consensus that future international action should include not only new emissions-reduction commitments by industrialized countries but also "nationally appropriate mitigation actions by developing country Parties . . . in a measurable, reportable and verifiable manner," and "policy approaches and positive incentives on issues relating to reducing emissions from deforestation and forest degradation in developing countries" (Bali Action Plan 2007).

This agreement shows that the world's governments, including the United States and large developing countries, realize that all major emission sources—industrialized countries, large developing countries, and deforestation—must eventually be included if climate mitigation is to succeed in the long run. The parties' "common but differentiated responsibilities" will be respected, and industrialized countries will be required to take far more action, much sooner and on many more types of emissions. Fortunately, a working framework and a timetable for a new agreement have been achieved. The challenge now is to complete work on the large number of critical details.

Summary and Conclusion

Global climate change is the result of an intensification of the natural greenhouse effect, caused by the release of heat-trapping gases from human activities including deforestation and the burning of fossil fuels. An anthropogenic increase in the greenhouse effect has altered the energy balance in the climate system, increasing global average temperatures. Warming has already been observed and is expected to increase in the coming century. The resulting climate change will produce a large number of mostly negative and potentially very harmful impacts. These effects include changes to many ecosystems, the oceans, the cryosphere, global and regional weather patterns, human health, and food security.

Our understanding of global climate change has developed steadily since the concept was first outlined nearly 200 years ago. However, only in the last two decades have serious efforts been made to develop policies to mitigate climate change. European countries lead these efforts (discussed in Chapter 3), but climate action is accelerating at state, regional, and national levels in the United States. The UNFCCC and the Kyoto Protocol stand at the center of international climate policy. The United States is not party to the Kyoto Protocol, which imposes binding emissions reductions commitments on the world's industrialized countries, but it is participating in negotiations aimed at creating a successor agreement.

Efforts to address climate change have encountered a number of challenges. These include scientific uncertainties, the complexity of the problem, debates concerning the potential economic costs or benefits of mitigation options, the influence of special interests, and the necessity of international cooperation. In many instances, these debates have made it more difficult for U.S. policy makers to reach consensus. Chapter 2 explores several of the most important problems and controversies in detail.

References

Aldy, J. E., S. Barrett, and R. N. Stavins. 2003. "Thirteen Plus One: A Comparison of Global Climate Policy Architectures." *Climate Policy* 3 (4) (September): 373–397.

American Institute of Physics (AIP). 2007. "Historical Overview." [Online information; retrieved 7/14/08.] www.aip.org/history/sloan/gcm/histoverview.html.

Arctic Climate Impact Assessment (ACIA). 2004. *Impacts of a Warming Climate.* Cambridge, UK: Cambridge University Press.

Bali Action Plan. 2007. "Report of the Conference of the Parties on Its Thirteenth Session, Held in Bali from 3 to 15 December 2007. Addendum. Part Two: Action Taken by the Conference of the Parties at Its Thirteenth Session." UN document FCCC/ CP/2007/6/Add.1*. [Online report; retrieved 8/14/08.] http://unfccc.int/resource/docs/2007/cop13/eng/06a01.pdf#page=3.

Ban, K. M. 2007. "A Climate Culprit in Darfur." *Washington Post,* June 16, A15.

Biello, D. 2004. "Ice Core Extends Climate Record Back 650,000 Years." [Online article; retrieved 7/14/08.] www.sciam.com/article.cfm?articleID=00020983-B238-1384-B23883414B7F0000.

Centers for Disease Control and Prevention 2008. "Malaria Facts." [Online information; retrieved 8/19/08.] http://www.cdc.gov/malaria/facts.htm.

Central Intelligence Agency (CIA). 2008. *World Factbook.* [Online information; retrieved 3/04/08.] https://www.cia.gov/library/publications/the-world-Fact Book/index.html.

Charlson, R. J. 2007. "A Lone Voice in the Greenhouse." *Nature* 448:254.

Chasek, P., D. Downie, and J. W. Brown. 2006. *Global Environmental Politics,* 4th ed. Boulder, CO: Westview Press.

Climate Change Science Program (CCSP). 2004. *Our Changing Planet: The US Climate Change Science Program for Fiscal Years 2004 and 2005.* [Online report; retrieved 7/14/08.] www.usgcrp.gov/usgcrp/Library/ocp2004-5/default.htm.

Dasgupta, S., Laplante, B., Meisner, C., Wheeler, D., and Yan, D.J. 2007. "The Impact of Sea Level Rise on Developing Countries: A Comparative Analysis." World Bank Policy Research Working Paper No. 4136. February 2007.

Delmas, R. J., J. M. Ascenio, and M. Legrand. 1980. "Polar Ice Evidence that Atmospheric CO_2 20,000 Yr BP Was 50 Percent of Present." *Nature* 284:155–157.

Dessler, A., and E. A. Parson. 2006. *The Science and Politics of Global Climate Change: A Guide to the Debate.* Cambridge, UK: Cambridge University Press.

Diouf, J. 2007. "Climate Change Likely to Increase Risk of Hunger." Speech presented at the M. S. Swaminathan Foundation Conference, Chennai, India, August 7.

Doney, S. 2007. "Effects of Climate Change and Ocean Acidification on Living Marine Resources." Written testimony presented to the U.S. Senate Committee on Commerce, Science and Transportation's Subcommittee on Oceans, Atmosphere, Fisheries, and Coast Guard, May 10. [Online testimony; retrieved 7/14/08.] www.whoi.edu/page.do?pid=8915&tid=282&cid=27206.

Downie, D., J. Gellers, and H. Raven. 2006. "Consensus Statements on Global Climate Change: An Updated Indicative Summary." Reference document submitted to the Fourth Meeting of the Global Roundtable on Climate Change, New York, December 18–19.

Energy Information Administration (EIA). 2008. Various official energy statistics from the U.S. Department of Energy. [Online information; retrieved 7/14/08.] www.eia.doe.gov/emeu/international/oilconsumption.html; www.eia.doe.gov/emeu/international/coalproduction.html; www.eia.doe.gov/dnav/pet/pet_cons_psup_dc_nus_mbbl_m.htm.

European Environment Agency (EEA). 2004. *Impacts of Europe's Changing Climate.* [Online report; retrieved 08/17/08.] http://reports.eea.europa.eu/climate_report_2_2004/en

Faris, S. "The Real Roots of Darfur." 2007. *The Atlantic Monthly.* April.

Fleming, J. R. 1998. *Historical Perspectives on Climate Change.* Oxford, UK: Oxford University Press.

Fourier, J.-B. J. 1827. "On the Temperatures of the Terrestrial Sphere and Interplanetary Space." *Mémoires de'l Académie Royales des Sciences de l'Institute de France,* trans. R. T. Pierrehumbert. [Online article; retrieved 7/14/08.] http://geosci.uchicago.edu/~rtp1/papers/Fourier1827Trans.pdf.

Gullison, R. E., P. Frumhoff, J. G. Canadell, C. B. Field, D. C. Nepstad, K. Hayhoe, R. Avissar, L. M. Curran, P. Friedlingstein, C. D. Jones, and C. Nobre. 2007. "Tropical Forests and Climate Policy." *Science* 316 (5827): 985–986.

Intergovernmental Panel on Climate Change Working Group I (IPCC WG I). 2007. *Climate Change 2007: The Physical Science Basis. Contribution of Working Group I to the Fourth Assessment Report,* edited by S. Solomon, D. Qin, M. Manning, Z. Chen, M. Marquis, K. Averyt, M. Tignor, and H. L. Miller. Cambridge, UK: Cambridge University Press. [Online information; retrieved 7/14/08] www.ipcc.ch/ipccreports/ar4-wg1.htm.

Intergovernmental Panel on Climate Change Working Group II (IPCC WG II). 2007. *Climate Change 2007: Impacts, Adaptation, and Vulnerability. Contribution of Working Group II to the Fourth Assessment Report,* edited by M. L. Parry, O. Caniziani, J. Palutikof, P. van der Linden, and C. Hanson. [Online information; retrieved 7/14/08.] www.ipcc.ch/ipccreports/ar4-wg2.htm.

Intergovernmental Panel on Climate Change Working Group III (IPCC WG III). 2007. *Climate Change 2007: Mitigation. Contribution of Working Group III to the Fourth Assessment Report,* edited by B. Metz, O. Davidson, P. Bosch, R. Dave, and L. Meyer. [Online information; retrieved 7/14/08.] www.ipcc.ch/ipccreports/ar4-wg3.htm.

International Pacific Halibut Commission (IPHC). 2007. "About IPHC." [Online information; retrieved 6/14/08.] www.iphc.washington.edu/halcom/about.htm.

Kane, S., M. Leiby, N. Paul, R. D. Perlack, C. Settle, J. F. Shogren, J. B. Smith, and T. J. Wilbanks. 2003. "Integrating Mitigation and Adaptation (Possible Responses to Global Climate Change)." *Environment* 45 (5). June 1, 28–39.

Kump, L. R., J. F. Kasting, and R. Crane. 2003. *The Earth System,* 2nd ed. Upper Saddle River, NJ: Prentice-Hall.

Lindsey, R. 2007. "Tropical Deforestation." [Online article; retrieved 7/14/08.] http://earthobservatory.nasa.gov/Library/Deforestation/deforestation_update.html.

Lorius, C., C. Ritz, L. Merlivat, N. I. Barkov, S. Korotkevich, and V. M. Kotlyakov. 1985. "A 150,000-Year Climatic Record from Antarctic Ice." *Nature* 316:591–596.

Mall, R. K., A. Gupta, R. Singh, and L. S. Rathore. 2006 "Water Resources and Climate Change: An Indian Perspective." *Current Perspective* 90 (12): 1610–1626.

National Aeronautics and Space Administration Earth Observatory (NASA EO). 2008. "The Carbon Cycle." [Online article; retrieved 7/14/08.] http://earthobservatory.nasa.gov/Library/CarbonCycle/.

National Snow and Ice Data Center (NSIDC). 2007. [Online information; retrieved 4/07/08.] http://nsidc.org.

Neumann, J. E., G. Yohe, and R. Nicholls. 2000. "Sea-level Rise and Global Climate Change: A Review of Impacts to U.S. Coasts." Washington, DC: Pew Center on Climate Change.

Pew Center on Global Climate Change. 2008. "What's Being Done . . . in the States." [Online information; retrieved 8/15/08.] www.pewclimate .org/what_s_being_done/in_the_states.

Revkin, A. 2007. "Agency Affirms Human Influence on Climate." *New York Times,* January 10, A16.

Schiermeier, Q. 2007. "The New Face of the Arctic." *Nature* 446 (7132): 133–135.

Tubiello, F. N., J. A. Amthor, K. Boote, M. Donatelli, W. E. Easterling, G. Fisher, R. Gifford, M. Howden, J. Reilly, and C. Rosenzweig. 2007. "Crop Response to Elevated CO_2 and World Food Supply." *European Journal of Agronomy* 26:215–223.

United Nations Department of Economic and Social Affairs (UNDESA). 2006. "World Population Prospects: The 2006 Revision Population Database." [Online article; retrieved 7/14/08.] http://esa.un.org/unpp/.

United Nations Food and Agricultural Organization (UN FAO). 2007. *State of the Forests 2007.* [Online report; retrieved 7/14/08.] www.fao .org/docrep/009/a0773e/a0773e00.HTM.

United Nations Framework Convention on Climate Change (UNFCCC). 1992. "United Nations Framework Convention on Climate Change." [Online report; retrieved 7/14/08.] http://unfccc.int/resource/docs/ convkp/conveng.pdf.

United Nations Framework Convention on Climate Change (UNFCCC) 2007a. "Kyoto Protocol: Negotiating the Protocol." [Online article; retrieved 7/14/08.] http://unfccc.int/kyoto_protocol/items/2830.php.

United Nations Framework Convention on Climate Change (UNFCCC) 2007b. "Land Use, Land-Use Change and Forestry." [Online article; retrieved 7/14/08.] http://unfccc.int/methods_and_science/lulucf/ items/4122.php.

U.S. Congress. 1997. "Byrd-Hagel Resolution." [Online U.S. Senate resolution; retrieved 7/14/08.] www.nationalcenter.org/KyotoSenate.html.

U.S. Department of Energy (US DOE). 2008. "How Fossil Fuels Were Formed." [Online information; retrieved 9/11/07.] www.fossil.energy .gov/education/energylessons/coal/gen_howformed.html.

U.S. House of Representatives, Committee on Oversight and Government Reform. 2007. "Committee Report: White House Engaged in Systematic Effort to Manipulate Climate Change Science." [Online report; retrieved 7/14/08.] http://oversight.house.gov/story.asp?ID=1653.

Vecchi, G. A., and B. J. Soden. 2007. "Global Warming and the Weakening of the Tropical Circulation." *Journal of Climate* 20:4326–4340.

Weart, Spencer R. 2004. *The Discovery of Global Warming*. Cambridge, MA: Harvard University Press.

World Bank. 2007a. "Agriculture and Rural Development—Deforestation." [Online information; retrieved 8/15/08.] http://web.worldbank.org/WBSITE/EXTERNAL/TOPICS/EXTARD/0,,contentMDK:20452548~pagePK:148956~piPK:216618~theSitePK:336682,00.html.

World Bank. 2007b. "Poverty at a Glance." [Online information; retrieved 8/15/08.] http://web.worldbank.org/WBSITE/EXTERNAL/COUNTRIES/LACEXT/EXTLACREGTOPPOVANA/0,,contentMDK:20564651~pagePK:34004173~piPK:34003707~theSitePK:841175,00.html.

World Energy Council. 2007. *Survey of Energy Resources*. [Online publication; retrieved 7/14/08.] www.worldenergy.org/publications/survey_of_energy_resources_2007/default.asp.

World Health Organization (WHO). 2007. "Health and Environmental Linkages: Climate Change." [Online information; retrieved 7/14/08.] www.who.int/heli/risks/climate/climatechange/en/.

Yeh, S. W., and B. P. Kirtman. 2007. "ENSO Amplitude Changes Due to Climate Change Projections in Different Coupled Models." *American Meteorological Society* 20:203–218.

2

Problems, Controversies, and Solutions

Introduction

In the United States, where skepticism of human-induced climate change once dominated government policy, a new consensus appears to have taken shape (Downie, Gellers, and Raven 2006). The vast majority of scientists agree that greenhouse gas emissions from human activity have begun to cause global climate change, that this pattern will accelerate without emissions controls, and that very serious impacts will result (IPCC 2007, 2001; USCSP 2006; G8 Science Academies 2005; Oreskes 2004). Hundreds of major corporations have publicly acknowledged the importance of the issue, taken internal mitigation measures, and/or called on governments to act (GROCC 2007; USCAP 2007; 3C 2007; Hoffman 2006). An increasing number of religious leaders from different faiths are affirming their belief that humans are changing the climate and that measures must be taken to reduce the scale and impact of this change (UUA 2006; Evangelical Climate Initiative 2006). News outlets give climate change topics prominent and consistent coverage, and books and documentaries on the topics proliferate. Thirty-seven states of the United States participate in regional climate initiatives, and more than half the states have developed specific action plans (Pew Center 2008). More than 850 mayors have signed the U.S. Conference of Mayors Climate Protection Agreement (U.S. Conference of Mayors 2008). A 2007 public opinion poll showed that 84 percent of Americans believed human activity was contributing to warming, with 80 percent believing immediate action was required (Broder and Connelly 2007).

Important skeptics remain, however. Sen. James Inhofe (R-OK) still questions the veracity of mainstream scientific claims and once called global warming the "greatest hoax ever perpetrated on the American people" (Inhofe 2003). President George W. Bush publicly doubted that the Earth's warming was man-made as late as 2006 and believes that mandatory emissions cuts are not required (Williams 2006). The opinion sections of several respected media outlets, such as the *Wall Street Journal*, continue to highlight the small number of scientists who dispute humanity's role in climate change or downplay the importance of the issue.

In addition, even though more and more Americans agree that anthropogenic warming is occurring, little agreement exists on what to do to stop it. Significant debates take place regarding how soon and how deep we should cut emissions, the best ways to do so, how to choose between various non-fossil-fuel technologies (wind, solar, geothermal, nuclear, wave, tidal, etc.), and how to best design, combine, or choose between myriad policy options. Many of these debates center on different cost and benefit views regarding climate change and potential policies. Economic uncertainties abound as well. A well-respected study calculates that the dire consequences of climate change can be avoided, with significant cost efficiency, for a price of about 1 percent of global gross domestic product (GDP) annually (e.g., Stern 2006). Others find that economic analysis justifies taking no major actions to mitigate climate change until far into the future (e.g., Nordhaus 2007). Still others argue that the economic costs of climate change cannot be calculated, as the consequences of climate change will be systemic and far reaching, meaning that even expensive mitigation actions should be taken immediately.

The contradictory claims regarding the extent of our climate knowledge, the viability of technological and policy options, and the economic impacts have often clouded the issue, confusing many experts and average citizens alike and impeding development of effective policy. To explore these issues, this chapter examines several of the most important and contentious problems, controversies, and the potential solutions relating to climate science, economics, and policy.

Scientific Controversies

In the last decade, some of the longest-running scientific debates about why and whether the climate is changing have given way to disputes over the magnitude and timing of the expected impacts. Which of the traditional controversies can we confidently say have been resolved? How can we evaluate arguments by the remaining skeptics? How does one evaluate competing claims? This section explores these issues, examining several of the classic problems and controversies of climate science, including a number of questions often raised by climate change skeptics. We begin by addressing whether the information to resolve these problems is available and/or reliable.

Is the Science of Climate Change Credible?

The Earth's climate system is extremely complex. This complexity and the fact that elements of the system can react nonlinearly or even chaotically (that is, randomly within set parameters) make it extremely difficult to predict next week's weather, let alone the Earth's overall climate in 50 years. Thus, some level of uncertainty is inevitable in climate science.

The presence of uncertainty does not mean that climate science is wrong, however, nor does it mean that people cannot act on the basis of its conclusions. It means only that scientists cannot determine with 100 percent accuracy what will occur in the future. This situation is not uncommon. For example, we know that smoking seriously impacts human health, causing heart disease, lung cancer, emphysema, and other ailments. The U.S. Centers for Disease Control and Prevention reports that smoking kills about 400,000 Americans each year. Nearly all scientists, government officials, and average Americans believe smoking is harmful. Yet science cannot predict with 100 percent certainty when a particular person will start feeling the most serious effects of smoking or which disease will become most severe. Nevertheless more and more young people choose not to start smoking because, even though they cannot predict exactly what damage smoking will cause to them and when, they know that "smoking kills." Some level of uncertainty is inevitable, but the science is credible.

The climate change situation is analogous but far more complex. Many problems and controversies in climate science have been resolved, including some discussed below, but uncertainty from a scientific perspective can never be eliminated. Even the most respected and agreed-upon scientific findings regarding climate change contain inherent unknowns that reflect what we do not yet know, or what is unknowable, about predicting the climate system. As a consequence, even honest scientific debates about particular uncertainties—current examples include climate sensitivity, the strength of aerosol forcing, the rate of ocean heat uptake, and tipping points—can make it appear that less is known about climate change than is actually the case.

Scientific uncertainty also makes it easy for those with vested economic or political interests to manipulate public perception of scientific knowledge and affect public policy discussion. As a result, the public sometimes finds it difficult to know which information is credible and which is less so. In the climate case, oil companies, the coal industry, and members of the second Bush administration have all, at different times, engaged in activities designed to dispute, obscure, or downplay the increasing scientific consensus regarding anthropogenic climate change and its impacts (U.S. House of Representatives 2007). Some energy companies, lobbyists, campaign contributors, and politicians have used perceived uncertainties to push for a "wait-and-see approach," urging the government to delay action until scientific evidence reduces the uncertainty regarding particular impacts. On the other side, some environmental groups have exaggerated evidence of climate change and levels of scientific knowledge concerning future impacts. Such situations are not uncommon and represent one reason that scientific complexity and uncertainty are well-known obstacles to creating effective national and international environmental policy (Axelrod, Downie, and Vig 2005; Chasek, Downie, and Brown 2006).

One way to reduce the negative impact of normal scientific uncertainty and to counter self-interested manipulation is to clearly and accurately convey the nature and degree of uncertainty associated with any particular finding. In this way, scientific findings are put into perspective, allowing people to judge information and related claims.

The standard method of conveying uncertainty in most scientific disciplines is probability. Probability calculations use real data to determine the statistical likelihood of certain events. For instance, the probability of droughts, storms, and floods has long been calculated with reference to historical precedent. However, because there is no historical precedent for anthropogenic climate change, statistical and probabilistic methods do not apply. As a result, climate scientists are forced to use subjective or qualitative descriptions of possible effects (Roe and Baker 2007). In addition, many of the scientific fields involved in interdisciplinary climate change research express the uncertainty associated with their research results differently. While some problems continue, improved understanding of climate change and more experience in conveying research findings to policy makers and the public have allowed researchers to develop new statistical, probabilistic, and qualitative descriptions of possible effects (e.g., those used in IPCC 2004; IPCC WG I 2007).

However, this acceptance does not solve the problem of what to do when competing claims are put forth regarding the uncertainties, nor when the amount of scientific evidence that must be considered is enormous. How do we judge competing claims? What do we do when data differ or when new claims are reported? How do we know when one opinion represents conclusions by dozens of the world's best experts while another is believed by only a few? How do we know media reports concerning science are accurate or that particular interests have not manipulated information?

Another method for reducing the negative impacts of scientific uncertainty is the creation of large, independent scientific panels to assess the best information available and report back to government officials and the general public. These bodies employ a review process that does not eliminate scientific uncertainty or competing claims, but does put such claims into perspective, allowing people to take action on the basis of the best consensus information available. Skeptics and alternative theories will always exist (a small number of people still claim that smoking is not harmful and that chlorofluorocarbons do not deplete the ozone layer), but over time, the controversies cease to impede reasonable action.

Such bodies have been employed as a way to address the climate problem, providing a solid basis to learn about the issue, evaluate uncertainties, and judge competing claims. Many governments, including that of the United States, have created special review panels or asked existing scientific bodies to develop consensus reports on key issues (e.g., USCSP 2006; NAS 2006). Perhaps the most important development occurred at the global level. During the mid-1980s, before international negotiations had begun and when far less climate science was known and agreed upon, many governments came to believe that they needed a credible, comprehensive source of information on climate change that would be trusted across national boundaries.

As a result, in 1987, the world's governments instructed two United Nations organizations, the World Meteorological Organization and the United Nations Environment Programme, to establish an intergovernmental body to assess the state of knowledge on global climate change. The goal of the resulting body, the Intergovernmental Panel on Climate Change (IPCC), is to provide the world's governments with periodic, comprehensive, authoritative, and up-to-date reviews of the scientific evidence regarding climate change so that governments have a scientifically sound and broadly accepted basis for making policy decisions (IPCC 2004).

The IPCC does not perform original research but rather performs exhaustive reviews on the current state of climate science. The process by which the IPCC develops its reports is designed to produce results that are credible within both the scientific community and the international political arena. First, the world's governments collectively elect the IPCC chair; nominate and elect the bureau (a group of distinguished experts that oversee the review process); and decide upon the general structure, mandate, and work plan of each assessment. The process is then turned over to the scientists.

Each major assessment (to date, assessments have been published in 1990, 1995, 2001, and 2007) is broken down into different subjects that are then examined by individual working groups (WGs). Each WG issues its own report. The first working group (WG I) examines the operation of the climate system and possible human impacts on that system that could lead to climate change (IPCC WG I 2007). WG II focuses on the expected impacts of climate change and possible adaptation to those impacts (IPCC WG II 2007). WG III examines policy and economic issues related to the mitigation of climate change (IPCC WG III 2007). A separate team then combines and integrates the major findings from each WG into a final synthesis report that provides an integrated picture of the climate issue (IPCC 2007). Governments can also request that the IPCC create special task forces to examine individual subjects as part of or between the major assessments (e.g., IPCC 2005).

The first draft of each IPCC report is written by dozens of the world's leading experts, who themselves often consult with a broader network of scientists. Once complete, this draft is sent to hundreds and sometimes thousands of other scientists around the world for review and comment before being revised. The final draft is then sent to governments for review and comment. Governments then approve the final document by consensus during a multiday meeting that sometimes includes intensive discussion and even negotiation. This final meeting, as well as previous sessions consisting of scientific discussions, is open to observers from research organizations, corporations, environmental groups, and universities.

To ensure that its work is based on the most reliable and credible information, the IPCC uses research that relies on the scientific method and peer-review process. Under the scientific method, an inquiry must begin with a hypothesis—an educated guess about how or why something works—that can be proved or disproved through observation and experimentation. The scientist predicts what the data will show and then gathers and analyzes data to determine whether or not it supports the hypothesis. If the hypothesis is supported by observations, the scientist submits his or her results for publication in a scientific journal. Before publication, other experts review the experimental design, the data collection techniques, and the results to judge whether the findings can be regarded as conclusive. The reviewers then recommend if the journal should accept or reject the paper, or suggest that the author revise and resubmit it. This process is called peer review.

Once a paper is published, other researchers attempt to produce the same results using other data or different techniques. A scientific claim that withstands many reviews, retests, and rigorous attempts to disprove it may, over time, become accepted as true. The IPCC's rigorous observance of this process contributes to its credibility. Moreover, because governments review and can propose changes to the final draft (including governments opposed to climate policy) and because timing issues exclude information circulating in the scientific community but not yet published, IPCC reports could even be considered conservative in their estimations of climate change and future impacts (Kerr 2007).

In December 2007, the IPCC shared the Nobel Peace Prize, one of the highest honors in the world, "for their efforts to build up and disseminate greater knowledge about man-made climate change." The 2007 assessment report is widely considered the most comprehensive, authoritative scientific statement on climate change that the scientific community has yet produced. Governments and scientists created the IPCC in part to address the types of questions listed at the start of this section. While uncertainties and skeptics will always exist, rigorous science; peer review; and global, comprehensive assessments have created a wealth of reliable, credible, and broadly accepted scientific information on climate change. With this background, we now turn to a series of key scientific questions, problems, and controversies.

How Do We Know the World Is Warming?

Though their numbers have dropped in recent years, a few vocal skeptics still believe no compelling evidence proves that the world is warming. Harvard University physicists Willie Soon and Sallie Baliunas, for instance, argue that the scientific evidence cannot prove the 20th century has been warmer than earlier historical periods (Soon and Baliunas 2003). Tim Ball, formerly of the University of Winnipeg, writes that the world's climate has actually cooled since 1940 (Ball 2004).

Despite these claims, evidence of global warming continues to mount. Indeed, research beginning in the late 19th century, conducted in many locations and by thousands of scientists, has led to a widespread consensus that the planet is currently warming at unprecedented rates, largely as the result of human activities. This section examines the scientific evidence that has led the vast majority of scientists to accept that the Earth is in fact warming.

Recent Temperature Trends

The most obvious place to look for evidence of warming is global temperature records. Thermometers have been in wide use for about 150 years. Records from thermometers around the world show that global average temperatures have increased over that time. According to these data, warming occurred in two distinct periods, 1910 to 1945 and 1976 to the present—with a cooling period in between. Overall, temperature records show an increase of 0.4 to 0.8°C during the 20th century, with roughly two-thirds of that warming occurring since 1978 (IPCC 2001, 2007).

Temperature is sometimes expressed in terms of temperature anomalies—that is, the difference between the temperature in a given year and some reference temperature, usually a long-term average. Evidence from the UK Meteorological Office shows that global average temperature anomalies have increased significantly in the past 50 years.

Of course, temperature data can include errors. If the time at which measurements are taken is not standardized, changes in temperature over the course of a day could affect calculations of average temperatures. Moving the thermometer from a shady to a sunny spot could have the same effect. Such sampling errors have occasionally been discovered, as the National Aeronautics and Space Administration did in some of its temperature data, and can generally be corrected using a suite of standard techniques (NASA 2007).

The distribution of measurement stations can also introduce potential errors. Most recordings are taken in areas of relatively dense population, with sparsely populated areas less well represented. Urban areas are, on average, warmer than nonurban areas. This "heat island" effect is primarily the result of concrete and asphalt. During the day, concrete and asphalt absorb and retain heat, warming the surrounding air more than natural vegetation would, particularly during the evening, as the concrete and asphalt cool down far more slowly. Scientists note the heat island effect could create a bias in temperature data—particularly because the number of urban areas, and urban temperature stations, has increased dramatically over the last 100 years. Recent studies demonstrate, however, that this effect has not impacted average annual temperature records and cannot account for observed warming (Parker 2004).

Beginning in the late 1970s, scientists have also used satellites to gather temperature data. Satellites do not measure temperature directly but collect data that can be used to infer surface temperatures. Interpretations of satellite data can vary widely, primarily because of the different mathematical assumptions that go into them, but the satellite data do show a warming trend (IPCC 2007; USCSP 2006).

The current consensus on the satellite data follows a period of debate during which some aspects of the data appeared to show that part of the lower atmosphere called the troposphere might be cooling. Atmospheric physics suggests that in the case of greenhouse warming, the troposphere should warm along with the Earth's surface. The possibility of a cooling troposphere was first reported in 1991 by two scientists at the University of Alabama, Huntsville (UAH), and was used by climate skeptics throughout the 1990s to cast doubt on climate change (Christy and Spencer 1990; Spencer and Christy 1992).

In 1995, three teams of scientists reexamined the UAH data. Each team found that the method used by the UAH team to compensate for the changing position of the satellites was flawed and had led to erroneously low temperature data. After all three teams published their results in the respected journal *Science* (Mears and Wentz 2005; Santer, Wigley, and Mears 2005; Sherwood, Lanzante, and Meyer 2005), the UAH scientists revised their methods. The revised results showed a strong warming trend in the lower atmosphere (although they still did not completely match the surface temperature increase), supporting theories about global warming (Spencer 2006).

In 2006, these findings were further corroborated by a U.S. Climate Change Science Program report, which completed a comprehensive review of ground and satellite temperatures. This government report reconciled ground and satellite temperature records and concluded that, since 1950, the data showed that the Earth's surface and lower and middle atmosphere had all warmed, while the upper atmosphere had cooled (USCSP 2006). Because these are the sorts of changes one would expect from an enhanced greenhouse effect, it resolved the earlier controversy and largely settled the temperature debate in the United States.

Proxy Data

In addition to examining temperature records from the relatively recent past, scientists study the Earth's climate in the distant past. This approach allows them to understand how the changes we are seeing now compare to both past climates and previous cases of climate change. Scientists conduct such studies by analyzing the Earth's natural recorders of climate history—including tree rings, coral reefs, ice cores, and ocean sediments. These sources are called proxies, because the evidence they provide about climate is indirect.

Trees grow out as well as up. This growth adds layers of wood to their perimeter, which appear as rings in horizontal cross section. In general, trees add one ring per year, which allows scientists to deduce not just the tree's approximate age but also its growth from year to year. The width of each ring reflects the relative growth conditions of that year. Because tree growth is affected by factors that include temperature, precipitation, and even wind, tree rings carry a great deal of climate information. As a result, dendroclimatologists—scientists who study the climate record preserved in trees—are able to use observations of the way these factors combine to affect tree ring growth in the present to make inferences about how temperature has changed over the lifetime of the tree (Jacoby and D'Arrigio 1997).

Similar to the readily viewable growth in trees, corals grow by adding an easily observable layer of calcium carbonate every year. Importantly, changes in the geochemical content of these layers provide clues about the climate. This is because the ratio of oxygen isotopes to elemental oxygen (O^{18}/O^{16}) present in coral changes based on how much of each is present in the air and in the ocean at the time that growth takes place. (An isotope is an atom of an element that contains a certain number of neutrons in addition to the unique number of protons that defines the element. For instance, oxygen is uniquely defined as having eight protons in its nucleus. The primary isotope of oxygen, O^{16}, has eight neutrons, whereas O^{18} has two additional neutrons, which cause it to behave differently than O^{16}.) An increase in the ratio occurs as sea surface temperature increases and ocean salinity decreases. These two factors are correlated because rising temperatures lead to melting ice, which decreases salinity. Analysis of this ratio therefore provides information about the range of temperatures a particular coral reef experienced in its lifetime.

Ice cores have also provided a great deal of well-respected evidence about past climates and the rate of previous climatic change (IPCC WG I 2007). As explained briefly in Chapter 1, glaciers and ice caps form by the accumulation and compaction of snow—though the top layers of these formations may melt and refreeze many times, the ice at the base stays frozen and may be hundreds of thousands of years old. Drilling deep into these formations, scientists extract long cylinders called ice cores. Layer by layer, these cores reveal changes in the climate—including temperature, precipitation, and the concentration of different gases in the atmosphere—over the entire period of time represented by the ice.

By comparing the ratio of oxygen and hydrogen isotopes, scientists are able to determine the relative temperature of different time periods. The hydrogen isotope deuterium, for instance, is found in higher quantities in ice formed during relatively warm periods. Scientists can also use the presence of ash and dust to obtain information about atmospheric moisture and volcanic eruptions, which affect climate by lowering the amount of solar radiation that reaches the Earth. Ice cores can also be analyzed directly for the presence of CO_2 and other greenhouse gases trapped in air bubbles in the ice.

The longest ice core ever recovered revealed that current levels of CO_2 and CH_4 are higher than they have been in 650,000 years. The core showed that CO_2 concentrations varied between 180 and 300 parts per million (ppm), but these concentrations took at least 800 years to increase to that extent. In contrast, human activities have raised CO_2 concentrations from 280 to 400 ppm in about 100 years (Biello 2005).

Analyzing and assembling data from proxy sources to create a credible picture of past climate is a difficult process and, like any scientific study, one that is open to differences of opinion and criticism. Perhaps it should then come as no surprise that the use of proxy data is at the root of one of the best-known climate change controversies.

In 1998, an article by climatologists Michael Mann, Raymond Bradley, and Malcolm Hughes appeared in the respected scientific journal *Nature* discussing long-term temperature trends. Another article followed expressing much the same results (Mann, Bradley, and Hughes 1999). Both articles included a graph (see Figure 6.7 in Chapter 6) of temperatures over the last millennium based on proxy data. The graph supported the premise that temperatures are now higher than they have been in the last 1,000 years and that the timing of the sharpest temperature increases (the last 200 years) corresponded to a period of increasing greenhouse gas emissions from human activity.

After it appeared in the IPCC's Third Assessment Report (IPCC 2001), the figure gained notoriety as the "hockey stick graph" (the moniker describes the shape of the temperature increase depicted). Appearing in articles around the world, the hockey stick was seen as "visually arresting scientific support for the contention that fossil-fuel emissions are the cause of higher temperatures" (*Wall Street Journal* 2005). Following the IPCC (2001) report, in 2003, Stephen McIntyre, a Toronto-based minerals consultant, and Ross McKitrick, an economist at Canada's University of Guelph, published an article questioning Mann, Bradley, and Hughes's results in the social science journal *Energy and Environment* (McIntyre and McKitrick 2003).

McIntyre and McKitrick argued that the temperature increase depicted in the hockey stick graph resulted from flawed methodology in the use of proxy data, including "collation errors, unjustifiable truncations of extrapolation of source data, obsolete data, geographical location errors, incorrect calculations of principal components, and other quality control defects" (McIntyre and McKitrick 2003, 751). In response, Mann and colleagues submitted a partial correction to *Nature,* acknowledging small errors but arguing that the overall outcome had not been affected.

Until this point, the controversy followed the standard pattern of scientific discourse: discovery, publication, attempts at replication, criticism, adjustment, republication, etc. The debate entered the political arena when McIntyre and McKitrick met with Senator Inhofe, shortly after which Rep. Joe Barton (R-TX) wrote to Michael Mann, demanding that he share all his data, methods, and associated information with critics and U.S. congressional staff (Eilperin 2005). While Mann considered the request, U.S. House of Representatives Science Committee Chairman Sherwood Boehlert (R-NY) asked Barton to withdraw what Boehlert called a "misguided and illegitimate investigation," arguing that the purpose of the investigation seemed to be "to intimidate scientists rather than to learn from them, and to substitute congressional political review for scientific review" (Eilperin 2005). Eventually, the respected U.S. National Academy of Science (NAS) was commissioned to review the original study.

The NAS report, released in 2006, rejected the claims of McIntyre and McKitrick and endorsed, with a few reservations, Mann and colleagues' work. The report specifically supported their conclusion that the latter half of the 20th century was the warmest period under consideration, but stated the authors should have better communicated the uncertainty surrounding the proxy data of the distant past. The NAS also took issue with the prominence given to the graph in the IPCC report. It believed that, given the uncertainty of proxy data and the fact that it was one of the first published studies of its kind, a more cautious approach would have been appropriate (NAS 2006).

In addition to confirming that proxy data provide evidence of increasing temperatures, the episode serves to underscore the credibility of climate change science. While the graph and its presentation were not perfect, Mann and colleagues' work passed initial peer review in a very respected journal. When criticized, it was thoroughly reviewed by the authors and improved upon. On the other hand, the work of McIntyre and McKitrick failed the peer-review process when they tried to publish a further critique in *Nature* (RealClimate 2004). Similarly, attempts to analyze, critique, and reproduce Mann and colleagues' results have led to adjustments and refinements of the technique, while attempts to reproduce the work of McIntyre and McKitrick have shown their original claims largely spurious (Rutherford et al. 2005; NAS 2006).

In conclusion, there is now widespread agreement—including the IPCC, NAS, U.S. government, and all national and international scientific bodies—that ground-based temperature records, satellite temperature records, and proxy data all confirm that the world is warming. Although potential exists for errors in ground-based, satellite, and proxy temperature records, the methods for discovering and correcting such errors are well known. Those errors that do occur will eventually be discovered and corrected through the peer-review process.

How Do We Know the Warming Is Anthropogenic?

Even after it became clear that the Earth was warming, it was difficult for some experts to believe that a system as large and complex as the Earth's climate could be vulnerable to human interference. In addition, because climate varies naturally, proving that observed changes were anthropogenic required separating the changes induced by human activity from those associated with natural variability. This effort has been a long process, and is still continuing, but it is central to understanding the drivers of global climate change.

Natural Variability

The Earth's climate fluctuates. The glaciers in mountainous areas of the United States, Canada, and Europe are remnants of vast ice sheets that covered much of the Northern Hemisphere 20,000 years ago. During the age of the dinosaurs, 100 million years ago, temperatures were about 10°C warmer than today (Crowley 1996). Indeed, scientists believe that temperatures were considerably warmer than they are today for roughly two-thirds of the last 400 million years.

Scientists attribute ancient climatic shifts to a number of different factors, including changes in the Earth's orbit, which affect the amount of sunlight reaching different places on the planet; a change in the sun's intensity; the internal variability of the climate system; volcanic eruptions, which emit aerosols that exert a cooling effect; and natural greenhouse gases, including water vapor and CO_2, which have a warming effect (Crowley 1996). In an attempt to understand current warming, scientists looked at these factors to determine if they might be at work now. In doing so, scientists have now eliminated all natural explanations for the current warming trend.

Observations play a big part in this process. For example, readings of sun intensity taken by satellite have been used to prove that solar radiation has not played a primary role in late 20th-century warming (Lockwood and Frohlich 2007). Scientists have also shown that variations in Earth's orbit occur so slowly that they could not be the cause of the recent rapid warming. It has also been proven that internal climate variability has not produced a change such as that which is currently being observed in at least the last 1,000 years (Jones and Mann 2004). In contrast, ice cores drilled in Antarctica show that CO_2 and temperature patterns strongly correlate for 150,000 years or more (Lorius et al. 1985; IPCC WG I 2007).

Climate models have also helped distinguish between natural and anthropogenic warming. They are important because, while some tests have been conducted in greenhouse facilities, scientists cannot run direct experiments to test the effect that greenhouse gases will have on the Earth's climate system. As a result, scientists use virtual laboratories—climate models called general circulation models (GCMs)—that allow ideas to be tested in a controlled manner. These complex computer programs, based on fundamental physical laws, simulate the Earth system and help predict how the climate will behave under different scenarios.

However, proving a GCM 100 percent correct would involve making a prediction and then waiting patiently for 50 or 100 years to see if the prediction comes true. In order to speed up the process, scientists run the models to see if they can "predict" what we know about past climate and to show that they are accurate enough to be useful for predicting future climates. They test how well the model predicts seasons, climate variability, and other aspects of historical periods for which good climate records exist. If a mismatch between the model's predictions and the known record occurs, the model is reexamined, adjusted, and improved (or eventually discarded). As such, the process by which climate models are developed is consistent with the scientific method.

General circulation models—and more recently, regional climate models—have improved rapidly over the last 20 years. They have successfully predicted a number of climatological features of the 20th century, including the warming of ocean surface waters (Cane et al. 1997), amplification of warming trends in the Arctic region (Serreze and Francis 2006), and an energy imbalance between incoming sunlight and outgoing infrared radiation (Hansen et al. 2005).

Climate models also show that most of the warming observed in the last half century is anthropogenic. Indeed, no model based entirely on natural variability—that is, models that do not include greenhouse gas emissions from human activities—produces the temperature changes observed in the last 50 years. When man-made CO_2 is included, however, model results closely resemble temperature observations over the last 100 years (IPCC 2007; USCSP 2006). This factor has given much more credence to anthropogenic warming.

Radiative Forcing

As outlined in Chapter 1, a change in the Earth's energy or radiation balance can lead to temperature changes on the Earth's surface. This type of change is referred to as a forcing. A positive forcing raises the net energy entering or being retained by the system. If the amount of energy leaving the system remains the same, surface temperature will increase. The opposite occurs with a negative forcing. Researchers determine the effects of various gases and other influences on the radiation balance through experimentation.

Scientists know that human activities raise the level of greenhouse gases in the atmosphere, and the radiative forcing of the gases being added. This knowledge enables them to predict how much the temperature should change based on the accumulation of these gases. When models incorporate radiative forcings, including both natural and human-induced effects, the simulated climate responds with a temperature increase close to what has been observed. This outcome provides strong evidence that human activities are causing most if not all of the current warming (IPCC WG I 2007).

In conclusion, there is a strong scientific consensus that the Earth is currently warming as a result of human activities. This consensus includes the IPCC; the United States, European Union (EU), and other governments; and the world's most respected national and international scientific bodies. Scientists have shown that the known causes of climate variability—including changes in the Earth's orbit, fluctuations in the sun's intensity, and natural changes in greenhouse gas levels—are not responsible for present temperature changes. General circulation models indicate that anthropogenic CO_2 matches well with the amount and timing of the rise in temperatures. Indeed, climate models show that most of the warming observed in the last half century is anthropogenic, and cannot account for the observed warming without including greenhouse gas emissions from human activities.

How Much Warming Will Occur?

Although there is widespread agreement that the world will warm, perhaps very significantly, scientists are not yet clear on exactly how sensitive the Earth's system is to changes in atmospheric CO_2. Models and observations both still yield broad distributions for how much the global mean temperature would increase should the amount of CO_2 in the atmosphere double from its preindustrial level (280 ppm). Mainstream estimates currently run between 2.5 and 4.0°C (IPCC 2007), while climate skeptics such as Richard Lindzen of the Massachusetts Institute of Technology argue that global average temperatures might rise as little as 0.5°C if CO_2 doubles (Lindzen and Giannitsis 2002).

Several factors contribute to this uncertainty. One is the time lag between changes in the radiation balance and a resulting change in the surface temperature. This lag occurs because the oceans absorb most of the additional energy trapped by greenhouse gases and thus delay the increase in surface temperature. However, no one knows precisely how long this time lag is. Another source of uncertainty is the degree to which aerosols may counteract the warming effects of CO_2, slowing the rise in temperatures. There are also feedbacks between the planet's response to climate change and climate change itself, which may amplify or dampen the progress of change (Kump, Kasting, and Crane 2004).

These factors make it difficult to determine the precise path that global warming will take, which, in turn, produces debate about the size of the threat posed by anthropogenic warming. Some scientists and other observers use this uncertainty to argue that climate change impacts may be small. A much larger number assert that while it is impossible to predict precisely how much warming will occur, models continue to affirm that it will be significant.

How Do We Know the Impacts of Warming Will Be Significant?

To assess the possible impacts of anthropogenic climate change, scientists use models that incorporate previously observed changes. They start by estimating a range of possible outcomes, which involves producing a series of scenarios that include different assumptions about demographic, social, economic, technological, and environmental developments. These scenarios are used to project greenhouse gas emissions and temperature rise, which are then incorporated into previously constructed models to predict the impacts on biological systems, freshwater resources, oceans, and human societies.

Some scientists and observers who accept that anthropogenic warming has occurred nevertheless doubt that it warrants significant concern (Deschenes and Greenstone 2006; Mendelsohn and Williams 2004; Tol 2002a). In general, these authors maintain that the negative impacts of climate change will be slight and that there is little sense in bearing high costs to prevent warming that is currently less than 1°C and not projected to be higher than 7°C. Some argue that climate change will, on the whole, bring net positive impacts.

As evidenced by the IPCC, a very large number of other scientists believe that climate change will bring increased temperatures, changing rainfall patterns, more frequent flooding and drought, melting glaciers, higher sea levels, expanded ranges for disease vectors, a large number of extinctions, and many other negative impacts (IPCC 2007; IPCC WG II 2007). These impacts will vary by region, but they will have a large number of consequential, negative impacts on human systems, including the availability of clean water, food security, agriculture, fisheries, human health, and coastal and far northern infrastructure, among many others. Poor communities will be particularly hard hit.

Parts of some regions will enjoy certain benefits, particularly the far north, which could see milder winters; longer growing seasons; ice-free shipping lanes; and new sites for mineral, oil, and gas extraction. However, the overwhelming scientific consensus is that the balance of climate change impacts, unless mitigated by reductions in greenhouse gas emissions, will be very negative. More information on the range of impacts can be found in the relevant sections of Chapters 1 and 3, in Figures 6.12 and 6.13 in Chapter 6, and in the relevant IPCC reports.

Do We Know How Quickly Climate Change Will Happen?

While the other questions in this section have largely been resolved, scientific consensus does not exist regarding how soon the most serious impacts, such as large rises in sea level, will occur. Change may occur slowly and gradually, or abruptly and drastically.

Studies of past climates reveal that the Earth system has the potential to shift abruptly. Ice core records reveal that one type of abrupt shift, named Dansgaard-Oeschger (D-O) events, precipitated temperature increases of more than 10°C within a few decades. Though evidence of D-O events exists, the dynamics of such events are not yet well understood. No one is sure, for instance, what the relevant threshold is for the oceanic shift that causes D-O events, nor is there a consensus if we are close to approaching it.

Another potential abrupt change would be the rapid, non-linear disintegration of the Greenland or West Antarctic ice sheets, which would raise sea levels dramatically and have other impacts due to the large influx of fresh water into the oceans. Concern also exists that melting permafrost in Russia, Canada, and Alaska could release huge amounts of methane into the atmosphere. While no one knows the precise implications of such dramatic shifts, the economic and ecological consequences would be enormous, as neither system is equipped to adapt rapidly to such dramatic change (Alley et al. 2003).

Scientists cannot yet assign precise probabilities to the variables associated with abrupt climate change, but understanding and attempting to avoid it are priority concerns for some scientists and policy makers. They cite the precautionary principle as support for taking strong mitigation actions, both to avoid the chance of abrupt change and to reduce the severity of the impacts that will still occur with gradual change.

While no internationally recognized definition exists, the essence of the precautionary principle is that a lack of certainty regarding a particular threat to humankind or the environment should not be used as a reason to postpone measures to prevent that threat. Its proponents, the most important being the EU, believe the precautionary principle should serve as a guide for creating national and international law. The EU used the precautionary principle to guide important recent policies on chemicals, genetically modified organisms, and consumer protection. The EU, joined by the Alliance of Small Island States (AOSIS), also strongly supports employing the precautionary principle to guide global climate policy. It argues that, given the tremendous potential dangers, uncertainties regarding the timing and full extent of specific impacts cannot be used as arguments against taking concerted, long-term national and international action to reduce greenhouse gas emissions.

Despite efforts to promote the precautionary principle, arguments concerning economic costs and benefits continue to dominate many discussions of climate change. Several problems and controversies exist at the center of these discussions. The next section outlines two of the most important issues.

Economic Controversies

While some scientific uncertainties persist, debates about why and whether the climate is changing have given way to disputes over what should be done to reduce that change. Much of this discussion focuses on very different views of various economic costs and benefits. Two overarching controversies dominate these discussions: Do we know how much climate change will cost (and if so, how much)? Do we know how much effective mitigation will cost (and if so, how much)?

Do We Know How Much Climate Change Will Cost?

Beginning with William Nordhaus's 1982 article, "How Fast Should We Graze the Global Commons?" economics has been an integral part of our efforts to understand and combat climate change. Economic analysis can help to frame the problem and options. By assigning monetary values to various types of potential action and inaction, economics helps citizens and policy makers compare sometimes quite disparate choices.

Most economic analyses start by assessing the costs associated with climate change. The task is not easy. Directly or indirectly, climate change will affect nearly all economic activity on Earth. Because some impacts will not be felt until far into the future, economists must engage in highly sophisticated guesswork. At the same time, they must also make judgments about the value today of costs or benefits occurring in the future. As a result, these analyses are highly complex and the results vary extensively. In general, economists begin their analysis by dividing the impacts of climate change into market and nonmarket damages (Goulder and Pizer 2008).

Market impacts include the welfare effects that result from changes in the price or quantity of marketed goods and services. For instance, rising temperatures and altered precipitation patterns are expected to affect agricultural production, forestry, and tourism, among other sectors. Rising sea levels may alter the price and availability of coastal land or affect industries whose infrastructure is currently located in coastal areas. Individuals may also suffer economic losses from more frequent extreme weather events or changes in the availability of insurance that will accompany these developments.

Simply adding up these impacts is a tall order, but to truly assess the change in human welfare, economists must also account for substitutions. That is, if conditions make it unprofitable to grow a certain crop, farmers may switch to a more suitable crop with no noticeable welfare effect. Similarly, industries that find little demand for their services in one area may sometimes easily shift areas, or services, to accommodate the new conditions. Of course, some of these substitutions will be expensive: Shifting from one crop to another may force changes throughout the food-processing industry, for instance. Others will be carried out with very little cost to producer or consumer. Because potential substitutions are essentially unlimited, and because measuring their direct and indirect costs is difficult, making concrete estimates of the welfare effects of market impacts is a complex process.

While calculating the damage associated with market impacts is difficult, determining the damage associated with nonmarket impacts is infinitely more so. Nonmarket impacts refer to those effects to things not exchanged in the market, and in the case of climate change may include effects such as ecosystem change, species loss, the spread of infectious diseases, and the loss of a way of life. Assessing the damage associated with these events involves assigning nonmarket goods and services a market price. But how do economists measure the cost of a loss of utility from a more inhospitable climate or from the loss of ecosystem services? How do they place a numeric value on species extinctions or the loss of traditional ways of life?

To assign a market value to these items, economists have developed a number of methods for assessing what people would be willing to pay for them. Although these methods are far from perfect, they can produce rough estimates that are useful within the context of cost-benefit analysis. These methods include "stated preference" techniques, which focus on interviews and other direct evidence. Other common methods include "revealed preference" approaches such as hedonic pricing, which looks at how people make decisions on where to live, or travel-cost methods, which use the distance that people are willing to travel to a given location to help determine the value that people put on it.

Economists must also find ways to incorporate the costs associated with the increased risks that climate change will bring. These damages may be felt as market impacts—for instance, the risk associated with coastal flooding and the increased incidence of extreme weather events may cause property values to decline in coastal areas where storms are common. Climate risk may also produce nonmarket damages. Companies known to emit large quantities of greenhouse gases may find their reputations tarnished as consumer attitudes change. In this case, firms may suffer financial setbacks due to the loss of reputation or they may decide to invest to avoid this outcome. Calculating the amount an entity will lose due to climate-related risks is difficult given that the specifics of some risks are, as of yet, unknown.

Economists also struggle to account for the costs of climate change across space and time. For instance, they must develop a system to compare costs felt in developed countries like the United States to those felt in the world's poorest countries. At present, damages are generally calculated in U.S. dollars. While this kind of standardization is necessary, it also creates important distortions. Specifically, the models have no way to reflect the relative magnitude of impacts felt in countries with very different perspectives or standards of living. This effect is particularly apparent with respect to nonmarket impacts, because the revealed and stated preferences of low-income people in developing countries seem to indicate that they value many things quite differently from their counterparts in more affluent countries.

Another problem arises in trying to account for changes over time. Indeed, a central issue in climate change economics involves the struggle to assign a present value to mitigation benefits that will occur in the future. Because greenhouse gases persist in the atmosphere, these benefits must be measured on a time scale much longer than a human life. But how much are those future benefits worth to us today? The discount rate is an economic tool used to compare costs and benefits that accrue at different times. It reflects the fact that, when asked to choose between immediate or delayed rewards, most people value a small immediate reward the same as a larger one later.

However, in the case of climate change, the long time frame means no precedent exists for the value we should choose for the discount rate. Moreover, the choice is not between a small reward now and a larger one later on, but between the relative costs to different generations, even several different generations. While a number of economists have argued that giving future generations less weight than the current generation is "ethically indefensible" (Stern 2006), others believe that weighting current and future generations equally leads to paradoxical and even nonsensical results. Introducing an artificially low discount rate, they believe, decreases the efficiency of mitigation policy (Nordhaus 1997).

Using detailed models to tally market and nonmarket damages and to account for substitution and the discount rate, economists produce a rough estimate of the costs of climate change. In most cases, this number is reported as a percentage reduction in the predicted global GDP. Models have improved significantly since the 1990s—both as a result of improving climate science and because economists have employed a more coherent approach to substitution—but the resulting cost estimates, even among the leading experts, still vary by large amounts.

For instance, a 2006 study commissioned by the UK government and chaired by Sir Nicholas Stern estimated the cost of climate change to be "at least 5 percent of global GDP each year, now and forever" and as high as a crippling 20 percent (Stern 2006). William Nordhaus, a pioneer of climate economics, puts the number at 3 percent in 2100 and close to 8 percent by 2200

(Nordhaus 2007). Richard Tol, another climate economist, estimates the cost range to be between –2.7 and 2.3 percent of global GDP (Tol 2002b). Addressing costs just in the United States, a 2008 study led by Frank Ackerman and Elizabeth Stanton concluded that without mitigation measures, the economic cost of climate change to the United States from increased hurricane damages, residential real estate losses due to sea-level rise, increased energy costs, and water supply costs would be $1.9 trillion (in 2008 dollars), or 1.8 percent of U.S. output, per year by 2100.

This variability persists when the costs of climate change are reported as the "social cost of carbon." The social cost of carbon is the marginal cost to society when one additional unit of carbon is released into the atmosphere (in general, the term *marginal cost* refers to the cost of one additional unit). It is expressed as the total of the net costs and benefits expected as a result of the additional emission. The social cost of carbon is sometimes also expressed as the price at which a carbon tax would need to be set to achieve an optimal result in which the marginal cost of carbon equals its marginal benefit.

In 2007, the IPCC reported that the average of 100 peer-reviewed estimates of the social cost of carbon was $43 per ton, with individual predictions ranging from $10 to $350 per ton (IPCC WG III 2007). That is, each ton of carbon (tC) emitted will cost the people on Earth $10 to $350 in climate impacts. (Greenhouse gas emissions totaled 45,000 megatons of CO_2-equivalents in 2004.) A 2002 report commissioned by the UK government produced a range of $7 to $154/tC (Clarkson and Deyes 2002). In 2005, Tol analyzed 103 estimates and reported the mode to be $2/tC, the median $14/tC, and the mean $93/tC.

What accounts for the difference? In short, calculating the costs of climate change involves a lot of creative thinking. Market damages are difficult to measure, especially when substitutions are considered, and nonmarket damages even more so. No agreement exists on the most appropriate discount rate, which determines the present value of the costs or benefits in the future. In addition, the global output against which these numbers are compared is also an estimate. At the same time, scientific uncertainty exists about the timing and range of climate change impacts. Thus, predicting the costs of climate change is still as much art as science.

Do We Know How Much Mitigation Will Cost?

To make cost estimates for the impacts of climate change useful from a policy perspective, we also need estimates of how much it will cost to lower emissions enough to avoid these impacts. To this end, economic analyses weigh mitigation costs against the value of avoiding climate-related damage. However, assessing the costs of mitigation remains as complicated as assessing the costs of climate change.

In general, two methods are used for assessing mitigation costs. "Bottom-up" energy-technology models try to assess costs by collecting details on specific energy processes and products. They focus on one or a few sectors and make no attempt to capture the effects of changes in the price of energy-intensive goods. "Top-down" models try to capture changes to the whole economy—tracing relationships among fuel costs, production methods, and consumer choices. More broadly based, top-down models tend to include much less detail on specific products or processes (Goulder and Pizer 2008).

Regardless of the model used, one of the most important factors in determining mitigation costs is the baseline. The baseline is an underlying assumption about the future of business-as-usual emissions. It is based on predictions about population growth, development, and the availability of energy resources. Because it defines the amount of greenhouse gas mitigation required, the baseline greatly influences cost assessments. The baseline is, however, inherently difficult to calculate—both because predictions extend precariously into the future and because many significant impacts are expected to occur in developing countries, where population growth and economic projections remain highly uncertain (Weyant 2000). For this reason, economic analyses can vary widely in terms of the baseline values they employ.

The potential positive effects of carbon abatement depend on the geographic scope of the mitigation effort under review. The costs of mitigation do as well. Will the mitigation under review take place only in the United States? In all industrialized countries? Or in industrialized countries plus large developing countries? The answers to these questions affect cost calculations. For example, developing-country participation opens up a wide range of low-cost mitigation opportunities not available in the developed world. Trade patterns are also important; countries that participate may find their costs determined in large part by the participation—or lack thereof—of others with which they trade climate-relevant goods (Tol 2006).

The long-term nature of the climate problem makes accounting for technological change another important part of cost assessments. Whether it occurs organically or as a result of specific policies, technological change will tend to lower the cost of greenhouse gas mitigation. The same is true for substitution. As mitigation policies make greenhouse-gas-intensive goods relatively more expensive, producers can be expected to substitute inputs or change to a cleaner product mix. The greater the potential for this kind of substitution, the lower the cost of mitigation. Yet assessing the potential for substitution and technological change remains complex and involves making assumptions about how costs will change in the long or short run.

Taking these factors into account, economists have developed broad estimates of mitigation costs. The Stern Report suggested that investing only 1 percent of global GDP annually in mitigation would avoid the worst effects of climate change (Stern 2006). Given the higher costs the Stern Report assigns to climate impacts, investing this amount would yield positive economic results. A report by the McKinsey consulting firm suggests that it would be possible to avoid 26.7 gigatons of carbon emissions for less than $60 per ton (Enkvist, Naucler, and Rosander 2007). A meta-analysis (that is, a study of different studies) conducted in 2003 reported that the range of costs of mitigation was between –3 percent and +2.5 percent of global GDP (Fischer and Morgenstern 2003). A 2008 meta-analysis by Yale University examined thousands of policy simulations from 25 leading economic models used to predict the economic impacts of reducing U.S. carbon emissions and concluded that even under the most pessimistic assumptions, reducing U.S. greenhouse gas emissions 40 percent by 2030 would still result in national GDP growth of 2.4 percent a year (Repetto 2008). Figures 6.17 and 6.18 in Chapter 6 provide more information on the costs of mitigation.

In conclusion, we do not know exactly how much climate change will cost nor how much climate mitigation will cost. A number of important estimates seem to indicate, however, that the economic benefits of avoiding the most serious climate change impacts outweigh the economic costs of reducing emissions. In this way, economics gives us the tools to make better-informed decisions about the costs and benefits associated with different courses of action. However, the complexity and uncertainty involved mean that cost estimates involve a lot of guesswork. As estimates improve, they will provide a better understanding of the economic trade-offs associated with addressing, or failing to address, climate change.

Policy Controversies and Solutions

Climate change affects many different political groups. Furthermore, the scale and scope of the problem, the range of policy options, and the potential for significant economic winners and losers combine to make discussions of climate policy difficult and divisive. This section outlines some of the major issues and controversies surrounding climate change solutions.

Climate Policy Basics

Emissions-reduction targets can be set either through absolute limits or intensity limits. Absolute limits assign a fixed cap on emissions from a country or particular sector; lowering the limit reduces emissions levels. Intensity limits, on the other hand, link emissions to another factor, commonly the size of the country's income or growth. Reducing the intensity limit reduces the emissions per unit of that factor; for instance, emissions per unit of GDP.

Absolute limits allow policy makers to know the maximum level of allowable emissions that will occur and to craft specific reduction schedules. Intensity limits focus on improving energy efficiency. They may be a useful step in limiting the growth of emissions, but they do not work well if the goal is to lower overall emissions. A mandate to reduce emissions per unit of GDP could be met even while overall emissions continue to rise.

Whether emissions limits are absolute or based on intensity targets, a number of strategies can be employed. The most common are conventional "command and control" regulations, market-based approaches, direct public expenditure, and voluntary initiatives. We will consider each strategy separately below, but it is important to keep in mind that many existing emissions-reduction policies, and nearly all proposals for future policies, contain elements of all four strategies.

Conventional regulation is what most people think of when they think of environmental policy. It involves the imposition of quantitative limits on pollution and is sometimes known as command and control. Command-and-control policy has been a popular and effective approach to environmental regulation in the United States and Europe since the 1970s, most notably in the areas of air and water pollution.

One reason that command-and-control regulations have been popular is because they allow regulators to set and enforce absolute limits on the total amount of pollutants being discharged. They may include the establishment of limits for pollutant levels in the surrounding environment (for example, the U.S. Clean Air Act sets ambient standard for six pollutants; if a region is in violation, the region must come up with a plan to achieve compliance) or the creation of emissions standards (for instance, the pounds per hour of sulfur dioxide that a plant can emit). An alternative form of command and control involves the enforcement of technology standards that require polluters to use certain technologies, practices, or techniques. An example of this is the requirement that automobiles driven in the United States use catalytic converters to reduce the toxicity of the emissions from internal combustion engines.

In contrast to command-and-control regulations, market-based approaches regulate emissions by creating economic incentives for emissions reduction. Many of these instruments, including emissions taxes and tradable emissions permits, are discussed in more detail in the next section. When they are well designed, market-based approaches can achieve similar levels of abatement as command-and-control regulations at much less cost. The United States successfully used emissions trading to reduce sulfur dioxide emissions from power plants. Partly based on this success, policy makers in the United States and elsewhere believe that emissions trading should be a part of any effort to reduce greenhouse gas emissions.

Direct public expenditure is when governments subsidize the use of existing technologies in the hopes that this will limit emissions. For instance, governments could subsidize the use of efficient lightbulbs, solar panels, or electric cars through tax incentives, rebates, or large purchases for government use. Government expenditure may also create favorable market conditions or reduce consumer risk. In this way, public expenditure does not discourage emissions so much as fund low-emission alternatives.

When the goal is to assist the development of new alternatives, public expenditure may go to research and development (R&D). Because R&D is expensive, private firms may not have the funds necessary to conduct it. If the society as a whole stands to benefit from a certain technological development, the government may choose to meet this funding gap. Governments can assist in the development and deployment of new technologies through grants, investments, contracts, tax incentives, targeted relaxation of antitrust laws, and other methods. While expenditure on R&D does not lead directly to short-term emissions reductions, it may speed up technological innovation that will lead to major reductions in the future.

Another strategy for reducing pollutants involves voluntary initiatives. These initiatives work to create favorable conditions for companies and individuals to take particular action. They may include public relations campaigns, public-private partnerships, or voluntary targets. Two voluntary initiatives currently underway in the United States are Energy Star, which is a voluntary labeling program designed to identify and promote energy-efficient products, and the Environmental Protection Agency's Climate Leaders initiative, which works with companies to develop strategies and goals for greenhouse gas reductions. Voluntary initiatives succeed when they create tangible incentives, such as labeling programs that impact consumer demand. Otherwise, nonparticipating companies can free-ride. Voluntary initiatives alone will have difficulty encouraging the scale of reductions necessary to address climate change.

Policy Options for the United States

As discussed in Chapter 1, the United States appears on its way to establishing a new climate policy in 2009 or 2010. The approved legislation will likely include binding targets for greenhouse gas emissions. Many hurdles remain, however, including choosing the specific mix of policy measures to be included. This section outlines some of the key solutions that the United States may choose to adopt in addressing climate change.

Cap and Trade or Carbon Tax?

Cap-and-trade policies limit greenhouse gas emissions by establishing a quantitative cap over the entire economy (or a specific sector, such as electric utilities). The total cap is then allocated to firms—through an auction or for free—in the form of permits, which give firms the right to a certain level of emissions. The emissions permits are then tradable between firms, allowing the cost of abatement to be equalized across the entire economy. Every year or so, the government then withdraws a certain number of permits, reducing the cap and, thus, emissions.

Designing a trading system can be tricky. The number of permits must be sufficient to allow trading and reductions to occur at a reasonable cost, but not so high as to eliminate the need for trading or reductions. The market must also be robust and transparent enough to provide firms with long-term confidence. In addition, verification and enforcement provisions must exist to ensure compliance, deter potential violators, and provide market confidence. The Kyoto Protocol, the EU, and several regional programs in the United States use cap-and-trade mechanisms.

In an emissions tax program, each firm pays the government a fee for each unit of pollution it emits (for example, each ton of CO_2). While firms with low abatement costs would reduce their emissions extensively in order to pay lower taxes, those with high abatement costs would reduce their emissions less and pay more in taxes. By establishing the cost at which firms decide whether it is more profitable to abate or pollute, a carbon tax sets the effective level of mitigation. The U.S. federal government and all state governments tax gasoline to raise transportation revenues, but this effort largely represents the extent of climate-related taxes in the United States.

In some ways, the approaches are similar. Both provide incentives for cost-effective mitigation and for innovation; both also increase energy prices and ultimately pass costs onto the consumer. Assuming the auction of cap-and-trade permits, both raise revenue as well.

The policies differ with regard to cost and emissions certainty. In the case of a tax, for instance, firms know how much the tax will be (cost certainty), but the level of emissions reductions that will be achieved across the country is uncertain (emissions uncertainty). Under a cap-and-trade program, governments set a cap on maximum emissions (emissions certainty), but firms do not know exactly what the permit price will be (cost uncertainty), a factor that might inhibit the market. One method to address cost uncertainty is to include a "safety valve" provision that would either cap potential permit prices or allow companies to opt out of the reduction requirements if they cannot purchase the necessary permits below a particular cost. In this case, however, the emission certainty of the trading scheme would be eliminated.

Trade-offs also exist with regard to R&D incentives. Under cap and trade, price volatility might scare businesses away from making long-term R&D investments. In a carbon-tax program, however, businesses would be confident of the price of carbon emissions—as a result they could easily incorporate expectations about energy prices into their long-term business plans. This approach might increase R&D investment.

Both systems present design difficulties. Choices are required on what emissions, industries, or activities will be subject to the tax or trade scheme; the breadth and depth of coverage; and implementation strategies, especially the rate at which taxes will increase (starting at the perceived maximum level immediately provides no preparation time for industry) or permits decline. Lobbyists and the congressional allies of all the potentially impacted industries will push for exemptions and loopholes. Provisions might be needed to ensure that goods entering the country face similar incentives to prevent emission leakage.

Once designed, tax systems are generally easier to administer; governments already possess revenue gathering and enforcement infrastructures. Administering trading schemes requires that exchange systems be created for the trades to take place and a verification mechanism to ensure that the traded reductions actually took place.

Carbon tax versus cap and trade is one of the most prominent climate policy debates. Both approaches have advantages and disadvantages. While both approaches have supporters in the United States, cap and trade appears to be winning. Only 2 of the 10 market-based climate change bills introduced in the 110th Congress (2007–2008) included taxes. In addition, although Senate Republicans blocked a formal vote on climate legislation in 2008, the failed legislation centered on a cap-and-trade mechanism, as will the likely successor attempt in 2009. However, future debates could be even more complicated as new support is gathering for sector-wide, as opposed to economy-wide, approaches (Climate Policy 2007). Sector-wide approaches would allow policies to treat sectors with long-lived infrastructure differently from those in which infrastructure and technology turned over relatively rapidly. On the other hand, it would make the regulation scheme much more complicated and leave more room for efforts to curry favor for particular sectors.

Nuclear

Currently, 104 nuclear reactors are operating in the United States, producing about 20 percent of the nation's electricity. The United States leads the world in electricity production from nuclear power, with about 97,000 megawatts. France, which produces 80 percent of its electricity from nuclear power, is second with about 63,000 megawatts; the EU as whole produces about 130,000 megawatts.

Interest in expanding nuclear power in the United States has grown significantly in recent years. Licensing applications are rising and significant building plans have been announced, although not initiated. This movement is a product of corporate economic interests; concerns for energy independence; the need to meet increasing energy demand while reducing greenhouse gas emissions; and the impact of federal policies aimed at assisting the industry, particularly the Nuclear Power 2010 Program and the 2005 Energy Policy Act. At the same time, opposition to building new nuclear plants remains high.

Nuclear energy is the most controversial alternative to fossil fuel power plants. The advantages of building new nuclear power plants include zero direct greenhouse gas emissions; no direct emissions of sulfur dioxide, nitrogen oxides, mercury, or other pollutants associated with burning fossil fuels; large, generally reliable facilities capable of base-load power plant operations; reduced dependence on oil and gas imports; improved designs that are approved and can be implemented immediately; and potential new plant designs that supporters claim will be less expensive to build and operate and produce less waste.

The disadvantages include the potential for accidental releases of radioactivity, including danger of a catastrophic accident; the potential for a terrorist strike designed to release a large amount of radioactivity; the production of highly radioactive waste requiring long-term storage; potential weapons proliferation through theft of waste or diffusion of nuclear technology and capability; high construction costs; the environmental impacts of uranium mining, including indirect greenhouse gas emissions; waste heat; and the economic risk that a multibillion-dollar facility could become nearly worthless in case of a significant accident.

It is impossible to conclude at this time if climate concerns combined with other issues will lead to a significant number of new nuclear plants. Opponents argue that energy efficiency, wind, solar, and geothermal can more than cover the perceived need for nuclear energy without the potential for catastrophe. This case demonstrates the trade-offs that need to be considered in order to address climate change. From this perspective, the debate over nuclear energy is also a climate controversy, and one that is unlikely to be resolved for many years.

Carbon Capture and Storage

The United States' current reliance on fossil fuels could make the transition to a more climate-friendly, low-carbon economy expensive and slow. Carbon capture and storage (CCS; also called carbon capture and sequestration) offers a way to speed up the process by reducing emissions while still using fossil fuels. In CCS systems, CO_2 created from the combustion of fossil fuels is captured before being released to the atmosphere and then stored in depleted oil and gas fields, unused coal seams, other geologic formations such as deep saline aquifers, or certain rocks with which the CO_2 can chemically combine, transferring carbon back to the geosphere. The CO_2 can be captured from any source but is most efficient from large stationary sources such as power plants, cement kilns, smelters, hydrogen production facilities, and other industrial plants.

Although CCS appears to be an easy solution to our carbon problem, several factors complicate its rapid introduction. While CCS technologies have been proven technically, they need to be demonstrated commercially and at the scale required to make a significant impact on efforts to decarbonize the global energy system. CCS requires significant amounts of energy and could actually increase our use of fossil fuels. CCS is also expensive. The IPCC estimates that CCS would increase the energy needs of power plants by 10 to 40 percent and the cost of energy from plants by 30 to 60 percent (IPCC 2005). Because of this cost, many companies have decided that CCS is not economically viable (Watson 2007). Moreover, to those who believe the goal of climate change mitigation should be to reduce our reliance on fossil fuels, CCS seems like a step in the wrong direction.

CCS also raises health issues. At high concentrations, CO_2 is toxic to breathe, so CO_2 escaping from storage sites would pose a potentially deadly health risk, depending on the concentration and duration of exposure. The accumulation of carbon dioxide in low-lying valleys would be a particular source of concern. Also, if stored CO_2 comes into contact with groundwater, the water might acidify. Another possibility is that groundwater might be contaminated by brine or metals displaced or mobilized by CO_2 injection.

The widespread deployment of CCS will also require resolution of a broad set of important legal, regulatory, and organizational issues. These include the industrial organization of carbon transport and storage, a variety of safety and integrity issues, the development of property rights for both geological formations and sequestered CO_2, immense liability issues (a key problem in the United States), and integration of CCS into carbon-market trading schemes. Until these issues are resolved, uncertainty will continue to impede large-scale investment and implementation of CCS.

Finally, geophysical limits could reduce the amount of CO_2 that can be stored. In 2005, the world emitted 27 gigatons of CO_2. After liquefaction, this amount would take up about 2.4 cubic miles—roughly the size of 2,500 football fields grouped together and extended skyward for two and a half miles (Kuo 2007). It seems likely that developing the infrastructure and locating the sites necessary to store all the carbon will be difficult, even if CCS systems are only applied to stationary sources with the largest emissions (Kuo 2007).

Geo-engineering

Geo-engineering involves the deliberate modification of the Earth's environment on a large enough scale to suit human needs. Until recently, such plans were the purview of science fiction. However, as the climate challenge escalates, more and more mainstream scientists have started to take geo-engineering seriously.

One such scientist is Paul Crutzen, corecipient of the 1995 Nobel Prize for chemistry for his research that helped alert the world to the possibility of ozone depletion. Crutzen believes that attempts to reduce greenhouse gases will fall so short of what is necessary that contingency plans are needed. He advocates releasing sulfur into the atmosphere to reflect sunlight before it reaches the planet (Crutzen 2006). This idea has gained support from some notable experts, including Wallace Broecker (Broad 2006; Caldeira 2007). Other geo-engineering ideas include using reflective films laid over deserts or white plastic islands floated on the oceans, both as ways to reflect more sunlight into space (Broad 2006), and adding relatively small amounts of iron to certain ocean regions to produce plankton blooms to consume carbon dioxide.

These and other proposals raise a number of important issues. Most central are the propriety and efficacy of intentionally altering the chemistry of the atmosphere or oceans in new ways. Advocates note that people are already altering the global environment and that these efforts will fix the problem. Critics argue that it appears safer to avoid or minimize global warming by reducing greenhouse gas emissions and restoring preexisting sinks, particularly forests. The possibility of unforeseen complications and unknowable side effects cannot be discounted, they say. The scale of such risks has produced opposition to many suggestions related to geo-engineering. As the risks of climate change mount, however, this attitude may change.

Mitigation or Adaptation?

Should the United States mitigate climate change by reducing greenhouse gas emissions or simply wait and adapt to the impacts of climate change? For many years, mitigation options dominated U.S. climate policy discussions, but recently, adaptation has become recognized as an equally important topic. Most experts now believe that both strategies should be pursued simultaneously and in coordination, and a great deal of work is being done in this area, although primarily outside the United States (Ingham, Ma, and Ulph 2005; Mills 2007; IPCC WG III 2007). For more information on adaptation, see Figures 6.15 and 6.16 in Chapter 6.

More broadly, some believe this question should be answered via cost-benefit analysis. Most economists would argue that policies should focus on mitigating greenhouse gas emissions only so long as the costs associated with climate damages exceed the cost of mitigation. When adaptation is taken into account, the calculus needs to include the costs and benefits of adapting to different levels of impacts. A number of challenges make it difficult, but not impossible, for economists to run cost-benefit analyses of the trade-off between adaptation and mitigation (Tol 2005).

As noted above, a small number of scientists, economists, and observers doubt that the impacts of the current warming trend will be significant and argue that taking expensive mitigation action is not warranted (Deschenes and Greenstone 2006; Mendelsohn and Williams 2004; Tol 2002a). By definition, this group would favor waiting to see the extent of the impacts and then adapting to them.

A far larger set of scientists, economists, policy makers, and business leaders foresee very large impacts and associate high economic costs with these impacts. In their view, mitigation is warranted by cost-benefit analysis, but so too are preparations to adapt to those changes that can no longer be avoided (e.g., IPCC WG III 2007; EC 2007; GROCC 2007; 3C 2007; Stern 2006).

Policy Options for the International Community

As noted in Chapter 1, international climate policy centers on the 1992 United Nations Framework Convention on Climate Change (UNFCCC) and its 1997 Kyoto Protocol. At present, the binding commitments of the Kyoto Protocol are not sufficient to adequately address the climate problem. This inadequacy is not surprising, as the protocol was always seen as the first step in a longer process, and in that regard it has succeeded. Nevertheless, it can be useful to outline the acknowledged failings of the Kyoto Protocol in order to understand the challenges for the global community in the years ahead.

Most basically, the industrialized countries will fail to meet the required emissions reductions. Instead of the mandated 5.2 percent reduction in emissions from 1990 levels, experts now predict about a 10 percent increase by the end of the commitment period in 2012 (e.g., Horner 2006). Nonparticipation by the United States prevents full compliance, but many other countries will also not meet their targets. Sweden, France, Germany, and the United Kingdom are on track to fulfill their role in achieving the 8 percent reduction required of the EU, but other European nations, including Austria, Denmark, Italy, Ireland, Portugal, and Spain, are off track. By 2004, the EU as a whole had only made a collective reduction of 0.9 percent (Horner 2006), although it still hopes to meet its requirement (EC 2007).

Japan is also struggling to meet its Kyoto targets (Suzuki 2006). Canada's emissions were 33 percent above its Kyoto commitment at the end of 2005, and it is widely agreed that they cannot be brought down to the required level within the compliance period (Dowd 2007). Australia and Norway are also not expected to meet their targets. Of course, these situations could change by extensive use of the protocol's flexibility mechanisms, the Clean Development Mechanism (CDM), Joint Implementation, and emissions trading, but it appears likely that the overall industrialized-country reduction target will not be met.

One reason for this collective failure is that Kyoto's commitments are, as many call them, "too little, too fast"—meaning that they are insufficient to do much about the climate change problem but still too ambitious for most countries in the short term. Also, nonparticipation by the United States made it politically easier for other countries to delay the policy adjustments necessary to meet their targets.

Another major problem with the Kyoto Protocol, at least from the perspective of mitigating climate change, is that it exempts many of the world's biggest emitters from the reduction targets. The exclusion of developing countries was a political necessity and upholds the important principle of "common but differentiated responsibilities" as set forth in the UNFCCC, but it makes it impossible to mitigate climate change effectively. By mid-2007, China had overtaken the United States as the world's largest greenhouse gas emitter, and emissions from India, Brazil, and other countries continue to increase (NEAA 2007). The increases in these countries easily dwarfs the cuts made in Europe.

Another issue is that the Kyoto Protocol does not effectively address emissions from land-use change. Deforestation, the draining of peat lands, and other activities account for more than 20 percent of annual global greenhouse gas emissions and represent the vast majority of emissions from many developing countries. For example, Indonesia is the world's fourth-largest emitter of greenhouse gases, and more than 80 percent of its emissions come from land-use change, especially forest clearing and burning (WRI 2007). In Brazil, 70 to 75 percent of greenhouse gas emissions come from deforestation and other land-use change (La Rovere and Pereira 2007).

The Kyoto Protocol also counts emissions reductions that have not and likely will not occur. The Kyoto Protocol uses 1990 as its baseline which means that all reduction targets refer to emissions levels in 1990. When the Soviet Union collapsed in 1991, its economy and its greenhouse gas emissions plummeted. Thus, Russia and the other countries that emerged from the breakup of the Soviet Union found themselves with "emissions reduction" targets well above their current level of emissions, which means they have far more emissions-trading permits than they need. When these permits are sold on the open market, other countries get credit for emissions reductions that are not actually taking place. These phantom reductions and their permits are commonly called "hot air."

As outlined in Chapter 1, countries are now engaged in negotiating a new agreement intended to take effect after the commitment period of the Kyoto Protocol expires in 2012. These negotiations began some years ago and are scheduled to conclude in December 2009, although it is far from clear that all the principal issues can be worked out by then.

Of the many challenges that must be overcome, the central one is getting all the major emitters and sources of emissions included, in some way, in the new agreement. This task will not be easy. The United States has consistently stated that it will not accept binding emissions reductions unless all other major emitters, including the largest developing countries, take on some type of commitments as well. This position receives quiet support from a number of other industrialized countries and, since it has great support in the U.S. Congress, it is unlikely to change after the 2008 elections. Developing countries meanwhile argue that their low per capita emissions and development needs justify them not making meaningful contributions to climate change mitigation until the developed world agrees to take on commitments for deep emission cuts, until the flexibility mechanisms and provisions for financial and technical assistance are improved significantly, and until the entire package is judged by developing countries as both effective and fair.

While this deadlock appears firm on the surface, signs are increasing, as outlined in Chapter 1, that an agreement can be reached. Some of the outlines of a potential agreement that run through various national proposals and interventions include industrialized countries accepting short-term reduction requirements (by around 2020–2025) and perhaps long-term reduction targets or goals (by 2050–2080), developing countries accepting a set of far less encompassing mitigation activities with provisions for increased developing country participation over time, the inclusion of avoided deforestation in the flexibility mechanisms, the expanded use of market-based mechanisms to lower the cost of mitigation, and provision of additional incentives for participation and compliance and, possibly, disincentives for nonparticipation (Aldy, Barrett, and Stavins 2003; ENB 2003–2008; personal observations by the authors).

Negotiations on these elements are moving so quickly that it would be fruitless to outline specific positions and proposals, as they could easily be resolved by the time this book is published. At the same time, circulating around the negotiations is a set of broader concepts and proposals designed to fix large-scale defects in the original protocol. Some of these are influencing the current negotiations and will likely remain relevant for some time. Others have little chance of receiving formal consideration in their current form but could still prove influential. To this end, the rest of this section briefly outlines these potential, and at times controversial, international solutions.

Avoided deforestation is widely recognized as one of the least expensive ways to reduce greenhouse gas emissions and would also greatly assist efforts to protect biodiversity. Consequently, a number of ideas have been proposed to protect forests from timbering or conversion to agricultural or pastoral uses. The potential solutions with the greatest long-term political potential center on efforts by rainforest nations and their nongovernmental organization allies to create new incentive mechanisms within the UNFCCC/Kyoto umbrella that would grant carbon credits for avoiding deforestation. This proposal found general agreement during the 2007 Bali climate negotiations. A policy that provides carbon credits within the emissions trading or CDM structures of the Kyoto Protocol could also deliver income to countries such as Brazil and Indonesia, create important economic incentives for landowners and the government, and provide an opportunity for developing countries with forest reserves to take on emission reductions as part of a new global policy structure. Regardless of the outcome of the 2009 negotiations, this issue will remain prominent for many years and debates on these types of proposed solutions will continue.

Many long-range proposed solutions center on expanding or otherwise improving emissions trading. One idea, first proposed early this decade, would allow international emissions trading without capping emissions. In this scenario, all nations—including developing countries—would be allocated permits equivalent to their anticipated business-as-usual emissions. These permits would be periodically retired by an international authority, which would offer to purchase them. Distributional issues would be resolved on the basis of per capita income levels and other criteria (Bradford 2002).

A related proposal would create a hybrid international trading instrument that combines an international trading mechanism, not unlike that found in the Kyoto Protocol, with a safety valve or price ceiling, to be implemented by an international agency making available additional permits at a fixed price. Proceeds from the sale of permits would finance climate change research and aid developing countries in reducing greenhouse gas emissions. Developing countries would also be included, at first via voluntary measures (Aldy, Orszag, and Stiglitz 2001). Both these potential solutions attempt to create true global emissions trading, reduce opposition to the program by limiting economic risk, and include developing countries.

Another set of proposed solutions would alter the traditional "target and timetable approach" used in many international regimes, including the Kyoto Protocol. The goals of these proposals are to attract broader participation in emission reductions, especially by the United States and the large developing countries, while encouraging more rapid conversion to clean technologies. For example, instead of emphasizing targets and timetables, one proposal emphasizes common incentives or requirements for climate-friendly technology. This approach could also employ a research and development protocol, which would be used to support collaborative research. Paired with this strategy would be a number of common-standards agreements, which would ensure that the fruits of such research were employed across the world. The proposal would focus on short-term progress without reliance on international enforcement (Barrett and Stavins 2003).

A related proposal also seeks to replace negotiations on national emissions quotas. In this proposal, countries would agree on a set of common actions aimed at achieving national and global emissions targets. These could include a harmonized carbon tax, integrated emissions trading, technology standards, renewable energy requirements, or other actions. This approach would allow for cost-effective mitigation but could also be negotiated in a more piecemeal fashion, bypassing the international climate-action log jam (Cooper 1998, 2001).

Other proposals in this area would keep targets and timetables but shift the focus to economic sectors. This approach could break negotiations into more manageable and potentially more productive individual agreements on emissions from, for example, power plants, transportation, landfills, buildings, agriculture, cement, steel, chemicals, and deforestation. This approach could also backfire, however: Breaking negotiations down in this way could lead to even greater complexity and opportunity for disagreement among countries on specific issues that might otherwise not arise in broader amalgamated climate negotiations.

Going further, another set of proposals would shift the focus away from a global accord entirely, moving outside the Kyoto process and concentrating on creating agreements within particular regions, among the largest emitters, or in particular economic sectors but outside the UNFCCC/Kyoto process. By removing the need to gather consensus from all countries, this type of agreement seeks to avoid the lowest-common-denominator problem and keep one or more laggard countries (e.g., the United States, Russia, or the Organization of the Petroleum Exporting Countries) from acting as a veto coalition and preventing an agreement (Chasek, Downie, and Brown 2006). President George W. Bush followed this path in 2007 and 2008 with his attempt to create a parallel negotiation track among the world's 16 largest economies, hoping that this set of large developed and developing countries, all with major greenhouse gas emissions, could reach consensus on a series of measures they could implement collectively.

Finally, some proposals would move entirely away from attempts to create binding international agreements and focus instead on voluntary actions, at least initially (Ansuategi and Escapa 2004). One proponent, President Bush, long advocated his view that the best course of action was for countries to pursue voluntary measures for reducing greenhouse gas emissions (Tollefson 2007). This stance angered European countries that believe that history shows voluntary targets do not work (MacAskill 2007). Some observers endorse this concept if the global negotiations become deadlocked. Rather than working fruitlessly to secure binding, global cooperation, these authors suggest that actions taken in individual countries, or agreed upon by small groups of countries, could serve as a model of behavior and make future cooperation easier. Once regulations to reduce greenhouse gases are common in many countries, the international community can come together to ratchet up commitments (e.g., Cashore et al. 2007).

Summary and Conclusion

The development of the climate change issue has been marked by a number of notable problems, uncertainties, and controversies. These include issues in climate science, the economic costs and benefits of climate change impacts and mitigation, and potential solutions. While several important scientific controversies have been resolved, no consensus exists on which potential solutions should be employed in the United States or globally. Several types of proven policy options exist, as do a large variety of proposed solutions, but questions remain as to whether the obstacles presented by competing economic interests can be resolved in the time frame necessary to avoid very serious impacts from global climate change. The answers to these questions lie not just in Washington but in the views and actions of several other countries around the world, each of which is arguably as important to the future of the global climate as the United States. Chapter 3 outlines these issues.

References

Achenbach, J. 2006. "The Tempest." *Washington Post*, May 28, W08.

Ackerman, F., and E. Stanton. 2008. "The Cost of Climate Change." [Online article; retrieved 7/15/08.] www.nrdc.org/globalWarming/cost/contents.asp.

Aldy, J. E., S. Barrett, and R. N. Stavins. 2003. "Thirteen Plus One: A Comparison of Global Climate Policy Architectures." *Climate Policy* 3 (4) (September): 373–397.

Aldy, J. E., P. Orszag, and J. Stiglitz. 2001. "Climate Change: An Agenda for Global Action." Paper prepared for The Timing of Climate Change Policies conference, Pew Center on Global Climate Change, the Westin Grand Hotel, Washington D.C., October 11–12, 2007.

Alley, R. B., J. J. Marotzke, W. D. Nordhaus, J. T. Overpeck, D. M. Peteet, R. A. Pielke, Jr., R. T. Pierrehumbert, P. B. Rhines, T. F. Stocker, L. D. Talley, and J. M. Wallace. 2003. "Abrupt Climate Change." *Science* 299 (5615): 2005–2010.

Ansuategi, A., and M. Escapa. 2004. "Is International Cooperation on Climate Change Good for the Environment?" *Economics Bulletin* 17 (7): 1–11.

Axelrod, R., D. Downie, and N. Vig. 2005. *The Global Environment: Institutions, Law, and Policy.* Washington, DC: CQ Press.

Ball, T. 2004. "Historical Climatologist: The Real Danger for Canada, Global Cooling." *Frontier Center for Public Policy,* November 15, 1.

Ball, T. 2007. "Global Warming: The Cold, Hard Facts?" [Online article; retrieved 7/15/08.] www.canadafreepress.com/2007/global-warming02 0507.htm.

Barnett, T. P., J. C. Adams, and D. P. Lettenmaier. 2005. "Potential Impacts of a Warming Climate on Water Availability in Snow-Dominated Regions." *Nature* 438 (17): 303–309.

Barrett, S., and R. Stavins. 2003. "Increasing Participation and Compliance in International Climate Change Agreements." *International Environmental Agreements: Politics, Law, and Economics* 3 (4): 349–376.

Biello, D. 2005. "Ice Core Extends Climate Record Back 650,000 Years." [Online article; retrieved 7/15/08.] www.sciam.com/article.cfm ?articleID=00020983-B238-1384-B23883414B7F0000.

Bradford, D. F. 2002. "Improving on Kyoto: Greenhouse Gas Control as the Purchase of a Global Public Good." Princeton University Working Paper, April 30.

Broad, W. J. 2006. "How to Cool a Planet (Maybe)." *New York Times,* June 27, F1.

Broder, J. M., and M. Connelly. 2007. "Public Remains Split on Response to Warming." *New York Times,* April 27, A20.

Caldeira, K. 2007. "How to Cool the Globe." *New York Times,* October 24, A19.

Cane, M. A., A. C. Clement, A. Kaplan, Y. Kushnir, D. Pozdnyakov, R. Seager, S. E. Zebiak, and R. Murtugudde. 1997. "Twentieth-Century Sea Surface Temperature Trends." *Science* 275 (5320): 957–960.

Cashore, B., G. Auld, S. Bernstein, and C. McDermott. 2007. "Can Non-State Governance 'Ratchet Up' Global Environmental Standards? Lessons from the Forest Sector." *Review of European Community & International Environmental Law* 16 (2): 158–172.

Chasek, P. S., D. L. Downie, and J. W. Brown. 2006. *Global Environmental Politics.* Boulder, CO: Westview Press.

Christy, J. R., and R. W. Spencer. 1990. "Precise Monitoring of Global Temperature Trends from Satellites." *Science* 247 (4950): 1558–1562.

Clarkson, R., and K. Deyes. 2002. "Estimating the Social Cost of Carbon Emissions." Working Paper 140. London: Public Enquiry Unit, HM Treasury.

Climate Policy. 2007. "Cap-and-Trade vs. Emissions Tax—Differences." [Online article; retrieved 7/15/08.] http://www.climatepolicy.org/?m =200708.

Combat Climate Change (3C). 2007. "3C—Combat Climate Change: A Business Leaders' Initiative." [Online information; retrieved 7/15/08.] http://www.combatclimatechange.org.

Cooper, R. 1998. "Toward a Real Treaty on Global Warming." *Foreign Affairs* 77 (2): 66–79.

Cooper, R. 2001. "The Kyoto Protocol: A Flawed Concept." *Environmental Law Reporter* 31:11484–11492.

Crowley, T. 1996. "Remembrance of Things Past: Greenhouse Lessons from the Geologic Record." *Consequences* 2 (1): 2–12.

Crutzen, P. 2006. "Albedo Enhancement by Stratospheric Sulfur Injections: A Contribution to Resolve a Policy Dilemma?" *Climatic Change* 77 (3–4): 211–219.

Deschenes, O., and M. Greenstone. 2006. "The Economic Impacts of Climate Change: Evidence from Agricultural Profits and Random Fluctuations in Weather." NBER Working Paper No. 10663, February. Cambridge, MA: National Bureau of Economic Research.

Dessai, S., and M. Hulme. 2005. "Does Climate Adaptation Policy Need Probabilities?" *Climate Policy* 4:107–128.

Dessai, S., X. Lu, and J. Risbey. 2005. "On the Role of Climate Scenarios for Adaptation Planning." *Global Environmental Change* 15 (2): 87–97.

Dessler, A.E., and E. A. Parson. 2006. *The Science and Politics of Global Climate Change.* Cambridge, UK: Cambridge University Press.

Dowd, A. 2007. "Canadian Province to Toughen Greenhouse Gas Rules." [Online article; retrieved 7/15/08.] http://uk.reuters.com/article/gc06/ idUKN1333848520070214.

Downie, D., J. Gellars, and H. Raven. 2006. "Consensus Statements on Global Climate Change: An Updated Indicative Summary." Reference document submitted to the Fourth Meeting of the Global Roundtable on Climate Change, New York, December 18–19.

Dume, B. 2007. "Making the Case for Carbon Storage." [Online article; retrieved 7/15/08.] http://environmentalresearchweb.org/cws/article/ opinion/31121.

DuPont, P. 2007. "Plus Ça (Climate) Change." *Wall Street Journal,* February 27. [Online article; retrieved 7/29/07.] www.opinionjournal .com/columnists/pdupont/?id=110009693.

Earth Negotiations Bulletin. "Climate Change Meetings." Coverage of Climate Meetings from 2003–2008. [Online information; retrieved 11/20/07.] www.iisd.ca/process/climate_atm.htm#climate.

Eilperin, J. 2004. "Humans May Double the Risk of Heat Waves." *Washington Post*, December 2, A10.

Eilperin, J. 2005. "GOP Chairmen Face Off on Global Warming: Public Tiff over Probe of Study Highlights Divide on Issue." *Washington Post*, July 18, A04.

Enkvist, P. A., R. Naucler, and J. Rosander. 2007. "A Cost Curve for Greenhouse Gas Reduction." *McKinsey Quarterly* February (Winter): 34–46.

European Commission (EC) 2007. "Limiting Global Climate Change to 2 degrees Celsius: The Way Ahead for 2020 and Beyond." [Online communication; retrieved 8/15/08.] http://eur-lex.europa.eu/LexUriServ/LexUriServ.do?uri=CELEX:52007DC0002:EN:NOT.

Evangelical Climate Initiative. 2006. "Statement of the Evangelical Climate Initiative." [Online communication; retrieved 8/15/08.] http://www.christiansandclimate.org/learn/call-to-action.

Fischer, C., and R. D. Morgenstern. 2003. "Carbon Abatement Costs: Why the Wide Range of Estimates?" Resources for the Future Discussion Paper 03-42, September. Washington, DC: Resources for the Future.

Global Roundtable on Climate Change (GROCC). "The Path to Sustainability: A Joint Statement by the Global Roundtable on Climate Change." February 20, 2007. [Online report; retrieved 8/09/07.] www.earth.columbia.edu/grocc/grocc4_statement.html.

Goulder, L. H., and W. A. Pizer. 2008. "The Economics of Climate Change." In *The New Palgrave Dictionary of Economics,* 2nd ed., edited by S. Durlauf and L. Blume. New York: Palgrave MacMillan.

Group of Eight (G8) Science Academies. 2005. "Joint Science Academies' Statement: Global Response to Climate Change." [Online information; retrieved 7/14/08.] http://www.scj.go.jp/ja/info/kohyo/pdf/kohyo-19-s1027.pdf.

Hansen, J., L. Nazarenko, R. Ruedy, M. Sato, J. Willis, A. Del Genio, D. Koch, A. Lacis, K. Lo, S. Menon, T. Novakov, J. Perlwitz, G. Russell, G. A. Schmidt, and N. Tausnev. 2005. "Earth's Energy Imbalance: Confirmation and Implications." *Science* 308 (5727): 1431–1435.

Herzog, H., and D. Golomb. "Carbon Capture and Storage from Fossil Fuel Use." In *Encyclopedia of Energy,* edited by C. Cleveland. New York: Elsevier.

Hoffman, Andrew. 2006. "Corporate Strategies That Address Climate Change." Arlington, VA: Pew Center on Global Climate Change.

Horner, C. C. 2006. "An Assessment of Kyoto and Emerging Issues for the 12th Conference of the Parties." *European Enterprise Institute. EEI Policy Note,* November 13, 1–25. http://ff.org/centers/csspp/library/co2weekly/20061201_02.pdf.

Ingham A., J. Ma, and A. Ulph. 2005. "How Do the Costs of Adaptation Affect Optimal Mitigation When There Is Uncertainty, Irreversibility, and Learning?" Tyndall Centre for Climate Change Research Working Paper 74. Norwich, UK: Tyndall Centre for Climate Change Research.

Inhofe, J. 2003. "The Science of Climate Change." [Online statement; retrieved 7/15/08.] http://inhofe.senate.gov/pressreleases/climate.htm.

Inhofe, J. 2006. "Inhofe to Blast Global Warming Media Coverage in Speech Today." [Online statement; retrieved 7/15/08.] http://inhofe.senate.gov/pressreleases/globalwarming.htm.

Intergovernmental Panel on Climate Change (IPCC). 2001. *Third Assessment Report: Climate Change 2001.* IPCC: Geneva.

Intergovernmental Panel on Climate Change (IPCC). 2004. "16 Years of Scientific Assessment in Support of the Climate Convention." [Online article; retrieved 7/15/08.] www.ipcc.ch/pdf/10th-anniversary/anniversary-brochure.pdf.

Intergovernmental Panel on Climate Change (IPCC). 2005. *Special Report on Carbon Dioxide Capture and Storage.* Cambridge, UK: Cambridge University Press.

Intergovernmental Panel on Climate Change (IPCC). 2007. *Climate Change 2007: The Synthesis Report.* Cambridge, UK: Cambridge University Press.

Intergovernmental Panel on Climate Change Working Group I (IPCC WG I). 2007. *Climate Change 2007: The Physical Science Basis. Contribution of Working Group I to the Fourth Assessment Report,* edited by S. Solomon, D. Qin, M. Manning, Z. Chen, M. Marquis, K. Averyt, M. Tignor, and H. L. Miller. Cambridge, UK: Cambridge University Press.

Intergovernmental Panel on Climate Change Working Group II (IPCC WG II). 2007. *Climate Change 2007: Impacts, Adaptation, and Vulnerability. Contribution of Working Group II to the Fourth Assessment Report,* edited by M. L. Parry, O. Caniziani, J. Palutikof, P. van der Linden, and C. Hanson. Cambridge, UK: Cambridge University Press.

Intergovernmental Panel on Climate Change Working Group III (IPCC WG III). 2007. *Climate Change 2007: Mitigation. Contribution of Working Group III to the Fourth Assessment Report,* edited by B. Metz, O. Davidson, P. Bosch, R. Dave, and L. Meyer. Cambridge, UK: Cambridge University Press.

International Pacific Halibut Commission. "About IPHC." [Online information; retrieved 7/15/08.] http://www.iphc.washington.edu/halcom/about.htm.

Jacoby, G. C., and R.D. D'Arrigio. 1997. "Tree Rings, Carbon Dioxide, and Climate Change." Colloquium Paper. *Proceedings of the National Academies of Sciences* 94:8350–8353.

Jones, P. D., and M. E. Mann. 2004. "Climate over Past Millennia." *Reviews of Geophysics* vol 42. RG 2002, 1–42.

Kameyama, Y. 2004. "The Future Climate Regime: A Regional Comparison of Proposals." *International Environmental Agreements* 4:307–326.

Kerr, R. A. 2007. "Climate Change: Global Warming Is Changing the World." *Science* 316:188–190.

Kump, L. R., J. F. Kasting, and R. G. Crane. 2004. *The Earth System.* Upper Saddle River, NJ: Prentice Hall.

Kuo, G. 2007. "Is Carbon Capture and Storage (CCS) Really the Technology to Save the Planet?" [Online article; retrieved 7/15/08.] www.amcips.org/articles/PDF/hansen_letter4-12-07.pdf.

La Rovere, E. L., and A. S. Pereira. 2007. "Brazil and Climate Change: A Country Profile." SciDev.net/en/policy-briefs/brazil-climate-change-a-country-profile.html.

Lackner, K. 2003. "A Guide to CO_2 Sequestration." *Science* 300 (5626), 1677–1678. June 13.

Lindzen, R. 2006. "Climate of Fear: Global-Warming Alarmists Intimidate Dissenting Scientists into Silence." *Wall Street Journal,* April 12, A14.

Lindzen, R. S., and C. Giannitsis. 2002. "Reconciling Observations of Global Temperature Change." *Geophysical Research Letters,* 29(12): 1583. 26 June.

Lockwood, M., and C. Frohlich. 2007. "Recent Oppositely Directed Trends in Solar Climate Forcings and the Global Mean Surface Air Temperature." *Proceedings of the Royal Society A* 463:2447–2460.

Lorius, C., C. Ritz, L. Merlivat, N. I. Barkov, S. Korotkevich, and V. M. Kotlyakov. 1985. "A 150,000-Year Climatic Record from Antarctic Ice." *Nature* 316:591–596.

MacAskill, E. 2007. "Europeans Angry after Bush Climate Speech 'Charade.'[ts]" *The Guardian,* September 29, 27.

Mann, M. E., R. S. Bradley, and M. K. Hughes. 1998. "Global-Scale Temperature Patterns and Climate Forcing over the Past Six Centuries." *Nature* 392 (6678): 779.

Mann, M. E., R. S. Bradley, and M. K. Hughes. 1999. "Northern Hemisphere Temperatures during the Past Millennium: Inferences, Uncertainties, and Limitations." *Geophysical Research Letters* 26 (6): 759–762.

McIntyre, S., and R. McKitrick. 2003. "Corrections to the Mann et al. (1998) Proxy Data Base and Northern Hemispheric Average Temperature Series." *Energy & Environment* 14 (6): 751–771.

Mears, C. A., and F. J. Wentz. 2005. "The Effect of Diurnal Correction on Satellite-Derived Lower Tropospheric Temperature." *Science* 309 (5740): 1548–1551.

Mendelsohn, R. O., A. Dinar, and L. Williams. 2006. "The Distributional Impact of Climate Change on Rich and Poor Countries." *Environmental and Developmental Economics* 11:159–178.

Mendelsohn, R. O., and L. Williams. 2004. "Comparing Forecasts of the Global Impacts of Climate Change." *Mitigation and Adaptation Strategies for Global Change* 9:315–333.

Mills, E. 2007. "The Role of US Insurance Negotiators in Responding to Climate Change." *UCLA Journal of Environmental Law and Policy* 26 (1): 129–168. December 22.

Mooney, C. 2004. "Beware 'Sound Science.' It's Doublespeak for Trouble." *Washington Post*, February 29, B02.

National Academy of Sciences (NAS). 2006. *Surface Temperature Reconstructions for the Last 2000 Years.* Washington, DC: National Academy of Sciences.

National Aeronautics and Space Administration (NASA). 2007. "GISS Surface Temperature Analysis 2007." [Online article; retrieved 7/15/08.] http://data.giss.nasa.gov/gistemp/updates/200708.html.

Netherlands Environmental Assessment Agency (NEAA). 2007. "China Now No. 1 in CO_2 Emissions; USA in Second Position." [Online article; retrieved 7/15/08.] www.mnp.nl/en/service/pressreleases/2007/20070619Chinanowno1inCO2emissionsUSAinsecondposition.html.

Nordhaus, W. D. 1982."How Fast Should We Graze the Global Commons?" *American Economic Review* 72 (2): 242–246.

Nordhaus, W. D. 1997. "Discounting in Economics and Climate Change: An Editorial Comment." *Climatic Change* 37 (2): 315–328.

Nordhaus, W. D. 2007. *The Challenge of Global Warming: Economic Policy and Environmental Policy.* [Online report; retrieved 7/15/08.] http://nordhaus.econ.yale.edu/dice_mss_072407_all.pdf.

Oreskes, Naomi. 2004. "The Scientific Consensus on Climate Change." *Science* 306 (3): 1686.

Parker, D. E. 2004. "Large-Scale Warming Is Not Urban." *Nature* 432:18.

Parry, I. W. H., and W. E. Oates. 2000. "Policy Analysis in the Presence of Distorting Taxes." *Journal of Policy Analysis and Management* 19 (4): 603–613.

Pew Center on Global Climate Change. 2008. "U.S. States & Regions." [Online information; retrieved 7/14/08.] www.pewclimate.org/states-regions.

Pielke, R., G. Prins, S. Rayner, and D. Sarewitz. 2007. "Lifting the Taboo on Adaptation." *Nature* 445:8.

Pizer, W. A. 2006. "Economics versus Climate Change." Resources for the Future. Discussion Paper 06-04. Washington, DC: Resources for the Future.

Real Climate. 2004. "False Claims by McIntyre and McKitrick Regarding the Mann et al. (1998) Reconstruction." [Online article; retrieved 7/15/08.] www.realclimate.org/index.php?p=8.

Real Climate. 2005a. "What Is a First-Order Climate Forcing?" [Online article; retrieved 7/15/08.] www.realclimate.org/index.php?p=186.

Real Climate. 2005b. "Is Climate Modelling Science?" [Online article; retrieved 7/15/08.] www.realclimate.org/index.php?p=100.

Repetto, R. "See for Yourself: How Reducing Greenhouse Gas Emissions Will Affect the American Economy." [Online article; retrieved 10/23/08.] An interactive Web site at www.climate.yale.edu/seeforyourself.

Roe, G., and M. Baker. 2007. "Why Is Climate Sensitivity So Unpredictable?" *Science* 318 (5850): 629–632.

Royal Society of Chemistry (RSC). 2006. "Can We Bury Our Carbon Dioxide Problem?" [Online article; retrieved 7/15/08.] http://www.rsc.org/ScienceAndTechnology/Policy/Bulletins/Issue3/CarbonDioxide.asp.

Rutherford, S., M. E. Mann, T. J. Osborn, R. S. Bradley, K. R. Briffa, M. K. Hughes, and P. D. Jones. 2005. "Proxy-based Northern Hemisphere Surface Temperature Reconstructions: Sensitivity to Methodology, Predictor Network, Target Season and Target Domain." *Journal of Climate* 18:2308–2329.

Santer, B. D., T. M. L. Wigley, and C. Mears. 2005. "Amplification of Surface Temperature Trends and Variability in the Tropical Atmosphere." *Science* 309 (5740): 1551–1556.

Serreze, M. C., and J. A. Francis. 2006. "The Arctic on the Fast Track of Change." *Weather* 61 (3): 65–69.

Sheppard, M. 2007. "Revised Temp Data Reduces Global Warming Fever." [Online article; retrieved 7/15/08.] www.americanthinker.com/blog/2007/08/revised_temp_data_reduces_glob.html.

Sherwood, S. C., J. R. Lanzante, and C. L. Meyer. 2005. "Radiosonde Daytime Biases and Late-20th Century Warming." *Science* 309:1556–1559.

Soon, W., and S. Baliunas. 2003. *Lessons and Limits of Climate History: Was the 20th Century Climate Unusual?* Washington, DC: George Marshall Institute.

Spencer, R. W., and J. R. Christy. 1992. "Precision and Radiosonde Validation of Satellite Gridpoint Temperature Anomalies. Part I: MSU Channel 2." *Journal of Climate* 5 (8): 847–857.

Spencer, R. W., J. R. Christy, W. D. Braswell, and W. B. Norris. 2006. "Estimation of Tropospheric Temperature Trends from MSU Channels 2 and 4." *Journal of Atmospheric and Oceanic Technology* 23 (3): 417–423.

Stangeland, A. 2006. "Carbon Capture and Storage—Technically Possible but Politically Difficult." [Online article; retrieved 8/15/08.] www.bellona.org/articles/articles_2006/ccs_technically%20possible_politically_difficult.

Stern, N. 2006. "The Stern Review Report." [Online report; retrieved 7/15/08.] http://www.hm-treasury.gov.uk./independent_reviews/stern_review_economics_climate_change/stern_review_report.cfm.

Suzuki, D. 2006. "Who's Meeting Their Kyoto Targets?" David Suzuki Foundation Backgrounder. Vancouver, BC: David Suzuki Foundation.

Tol, R. S. J. 2002a. "New Estimates of the Damage Costs of Climate Change, Part I: Benchmark Estimates." *Environmental and Resource Economics* 21 (1): 47–73.

Tol, R. S. J. 2002b. "New Estimates of the Damage Costs of Climate Change, Part II: Dynamic Estimates." *Environmental and Resource Economics* 21 (1): 135–160.

Tol, R. S. J. 2005. "The Marginal Damage Costs of Carbon Dioxide Emissions: Assessment of the Uncertainties." *Energy Policy* 33 (16): 2064–2074.

Tol, R. S. J. 2005. "Emission Abatement versus Development as Strategies to Reduce Vulnerability to Climate Change: An Application of FUND." *Environment and Development Economics* 10 (5): 615.

Tollefson, J. 2007. "Cool Reaction to Bush's Climate Summit: Emphasis on Technology over Emissions Targets Finds Little Favor." *Nature* 449:519.

Unitarian Universalist Association of Congregations (UUA). 2006. "Threat of Global Warming/Climate Change: 2006 Statement of Conscience." [Online statement; retrieved 6/6/08.] http://www.uua.org/socialjustice/socialjustice/statements/8061.shtml.

United Nations Development Program (UNDP). *Human Development Report 2007: Fighting Climate Change: Human Solidarity in a Divided World.* New York: UNDP.

United States Climate Action Partnership (USCAP) 2007. *A Call the Action.* Washington, DC: USCAP.

U.S. Climate Change Science Program (USCSP). 2006. *Temperature Trends in the Lower Atmosphere: Steps for Understanding and Reconciling Differences.* [Online report; retrieved 7/15/08.] http://www.climatescience.gov/Library/sap/sap1-1/finalreport/default.htm.

U.S. Conference of Mayors. 2008. "Mayors Climate Protection Center." [Online information; retrieved 7/14/08.] www.usmayors.org/climate-protection/.

U.S. House of Representatives, Committee on Oversight and Government Reform. 2007. "Committee Report: White House Engaged in Systematic Effort to Manipulate Climate Change Science." [Online report; retrieved 7/15/08.] http://oversight.house.gov/story.asp?ID=1653.

Wall Street Journal. 2005. "Hockey Stick on Ice—Politicizing the Science of Global Warming." Editorial. *Wall Street Journal,* February 18, A10.

Watson, C. 2007. "Statoil and Shell Decide against Carbon Capture Project." *Energy Business Review,* July 2. [Online article; retrieved 10/23/08.] www.energy-business-review.com/article_news.asp?guid=93A4AF76-7450-4BE1-A08D-43F82D75DBAA.

Weart, S. 2004. *The Discovery of Global Warming.* Cambridge, MA: Harvard University Press.

Weyant, J. 2000. "An Introduction to the Economics of Climate Change Policy." Arlington, VA: Pew Center on Global Climate Change.

Willett, K. M., N. P. Gillett, P. D. Jones, and P. W. Thorne. 2007. "Attribution of Observed Surface Humidity Changes to Human Influence." *Nature* 449:710–712.

Williams, C. 2006. "Bush Denies Human-Induced Climate Change. *The Register,* March 31. [Online article; retrieved 10/23/08.] www.theregister.co.uk/2006/03/31/bush_climate_logic/

World Resources Institute (WRI). 2007. "Painting the Picture of Tropical Tree Cover Change: Indonesia." [Online information; retrieved 7/15/08.] www.wri.org/biodiv/topic_content.cfm?cid=4526.

3

Worldwide Perspective

Introduction

Climate change is a global phenomenon. Temperatures will rise, rainfall patterns will change, and sea levels will increase around the world. The implications of these changes will reach across national boundaries and into future generations, affecting social, economic, and natural systems for years to come. But while climate impacts will affect all life on Earth, the scale and nature of these impacts will be neither uniform nor fair.

A variety of climatic, geographic, economic, and social factors make some countries and regions more vulnerable to climate change. Some countries even stand to benefit, at least initially. Expected impacts greatly influence the policies of some governments. For others, economic considerations are most important.

Chapters 1 and 2 include discussions of climate issues in the United States and the international policy process. This chapter outlines how climate change will impact other important countries and regions of the world; describes the policy perspective of these countries; and, where appropriate, draws a connection between the two. Unless otherwise noted, information in this chapter regarding current and expected impacts of climate change is drawn from reports of the Intergovernmental Panel on Climate Change (IPCC) (IPCC 2007; IPCC WG II 2007). The chapter begins by examining the situation of the world's most vulnerable countries, the small island developing states (SIDS) and nations in Africa, and then discusses four critically important and rapidly industrializing countries: China, India, Brazil, and Russia; the leader in mitigation efforts, the European Union (EU); and a set of countries with somewhat unique interests in the climate issue, the nations of the Organization of the Petroleum Exporting Countries (OPEC).

Small Island States

Fifty-one countries are classified as small island developing states (SIDS). Although they collectively account for less than 1 percent of greenhouse gas emissions, SIDS face climate change impacts that are dire and potentially disastrous. As a result, the Alliance of Small Island States (AOSIS), an international alliance created by SIDS, works to elevate climate change mitigation and adaptation on the international agenda.

The critical issue for most SIDS is sea level rise. The highest point in Tuvalu, for example, is just 5 meters above sea level, meaning that rising sea levels threaten to submerge most of the country within a generation (Price 2002). The Marshall Islands, the Maldives, Kiribati, the Netherlands Antilles, and Vanuatu could also disappear or see the vast majority of their territories become submerged. Other SIDS, including many well-known countries in the Caribbean and Pacific, could lose significant portions of their most populated and economically important coastal areas, dislocating millions of people.

Climate change will also reduce the resilience of coastal ecosystems, increase storm surges and coastal erosion, and lead to further saltwater intrusion into freshwater resources. Buildings and other infrastructure will be at greater risks from floods, landslides, and wind damage. Combined, these impacts will significantly affect SIDS's most important economic sectors, including fishing, agriculture, and tourism (Tompkins 2005). Coral reefs, important ecological systems which function as natural breakwaters and contribute to local economies through food production and ecotourism, also face several threats, including ocean acidification as a result of increased levels of CO_2, rising ocean temperatures, silting and other impacts of increased coastal erosion, and stronger storms.

Climate change will affect public health as well. Rising temperatures will likely increase the incidence and geographic distribution of malaria, dengue fever, filariasis, and schistosomiasis (Chen 2006). The disruption of sewage and water supply systems resulting from rising seas and stronger storms could increase the incidence of water-borne diseases as well. These issues, combined with the wide array of health risks associated with population displacement, could overwhelm the already strained health care infrastructures of most SIDS.

Many small islands also suffer from "double exposure"—geographic vulnerability to climate change coupled with poverty, underdevelopment, and a weak position in the global economy. Climate change will increase the vulnerability that these states already face from generally poor socioeconomic conditions; in turn, socioeconomic challenges will exacerbate the impacts of climate change (Kelly and Adger 2000). The ability of SIDS to adapt to, or cope with, climate impacts is constrained by their weak economies. In this sense, climate vulnerability is a product not just of geography but also of socioeconomic trends (O'Brien and Leichenko 2003).

Given their high level of risk, small island states have taken measures to increase their ability to adapt to climate change and otherwise reduce their climate vulnerability (Adger et al. 2003; Tompkins 2005). Most have established permanent national climate change focal points within their governments and have begun developing rapid-response contingency plans. Many are upgrading data collection systems, enabling them to gather, analyze, and interpret meteorological data, as well as participating in multinational programs such as the South Pacific Sea Level and Climate Monitoring Project.

SIDS are also engaged in several international adaptation planning programs, including Caribbean Planning for Adaptation to Climate Change and Mainstreaming Adaptation to Climate Change, which receive additional resources from donor nations and international organizations such as the World Bank and Global Environment Facility (GEF). The programs aim to strengthen the capacity of institutions in small island states to identify climate risks, prioritize risk-reduction opportunities, and begin taking appropriate action. Informing and educating the public to build the capacity of individuals to respond to changing conditions is another critical goal of these programs (Mathur 2001). Other national and international efforts seek to improve estimates of the economic value of coastal and marine resources in order to buttress arguments for their protection and attract funding (CARICOM 2008).

In addition to beginning to prepare for the impacts of climate change, SIDS also work to increase international action on climate mitigation and adaptation. This effort includes urging industrialized nations and large developing countries to reduce their greenhouse gas emissions, increase other mitigation efforts, and help SIDS receive the technical and financial assistance they need to prepare for, and adapt to, the impacts of climate change that can no longer be avoided.

Many of these efforts are done collectively through AOSIS, which functions as the negotiating voice for SIDS within the United Nations (UN) system. This work began even prior to negotiation of the United Nations Framework Convention on Climate Change (UNFCCC) but accelerated considerably afterward. Of particular importance was the 1994 Barbados Program of Action, in which 111 countries agreed upon priority areas for the sustainable development of small island states. Top among these priorities were addressing climate change and consequential sea level rise. In subsequent international forums, AOSIS has repeatedly voiced its distress at the world's collective failure to reduce greenhouse gas emissions, urging strong action. AOSIS has also proposed the development of a framework for industrialized countries to pay reparations for the climate-related damages that SIDS will experience as a result of climate change, including loss of territory, economic damages, and rising health care needs.

AOSIS consistently advocates significant strengthening of global controls on greenhouse gas emissions. For example, as early as 1995, AOSIS submitted a proposal to the first UNFCCC Conference of the Parties calling for industrialized countries to reduce their emissions of CO_2 by 20 percent from 1990 levels by 2005, a far stronger proposal than either the Kyoto Protocol or the current domestic policies of nearly every country outside the EU. SIDS also argue that climate change represents a threat to their citizens' fundamental right to both national sovereignty and a safe, secure, and sustainable environment. By framing climate change as a human rights issue, AOSIS hopes to increase attention to its members' plight and convince industrialized countries to address the severe climate risks being faced by small island states (Makan 2007).

To date, the international community has not agreed to make the kinds of emissions reductions necessary to safeguard SIDS against the worst impacts of climate change. As a result, some island states are faced with the possibility that their countries will become uninhabitable. Such an outcome would challenge not only the residents of small island states but also the international system itself. No precedent exists for the mass migration or loss of sovereignty associated with the physical disappearance of a nation. How the international community will respond to this problem remains to be seen.

Africa

Africa's 11.7 million square miles of territory account for 20 percent of the world's land area. Its 930 million human inhabitants live in 53 countries (islands included) and represent an extraordinary variety of ethnic, cultural, lingual, economic, and religious groups. Possessing immense cities and tens of thousands of small villages, Africa's diversity is also present in its physical geography and climate: The only continent to stretch from the northern temperate to the southern temperate zones, Africa is home to the world's largest desert, tall mountains, dense tropical rainforests, immense savanna plains, and vast coastlines.

Given such geographic and human diversity, the impacts of climate change on Africa will vary significantly. And yet, for Africans faced with a changing climate, the primary factor in determining their ability to deal effectively with changing conditions may be neither geographic nor cultural, but rather economic. Indeed, while some African countries have pockets of significant wealth, most of the continent is extremely poor. The United Nations classifies 33 African countries as among the 50 least developed countries in the world (UN 2007). As many as 400 million people live on less than a dollar a day (World Bank 2004). At least 40 million do not have enough to eat on a daily basis, far more have no regular access to clean drinking water, and millions are considered at risk of starvation without external assistance (WFP 2008). These economic conditions are one reason that Africa contributes only 5 percent of global greenhouse gas emissions.

These economic challenges severely constrain efforts to cope with climate change, providing almost no buffer for the impacts to come. Moreover, to those struggling to provide or access adequate health care and sufficient daily supplies of water and food, to curb civil and intrastate violence, and to improve basic infrastructure, climate change may seem a very remote problem.

Nevertheless, IPCC reports and other research indicate that Africa will experience significant, varied, and far-reaching impacts from global climate change. While climates across Africa have always been erratic, scientific research indicates the potential for new and dangerous extremes brought on by climate change (Magrath and Simms 2006). Temperatures will rise, rainfall patterns will change, and extreme weather events will increase in frequency and intensity, all of which will have impacts for agriculture and livestock. Human health and welfare will also decline, as the intensity and range of weather-modulated and/or mosquito- and pest-borne diseases, including malaria, yellow fever, cholera, and dengue fever, expand in many areas (Patz et al. 2005).

Some impacts are already occurring. On average, the African continent is 0.5°C warmer than it was 100 years ago. In some areas—including parts of Kenya where temperatures have risen 3.5°C in the last 20 years—changes have been much more dramatic. Some already arid or semi-arid areas are becoming increasingly drier, while the wet regions of equatorial Africa are getting wetter (Magrath and Simms 2006). The snow and ice atop Mount Kenya and Mount Kilimanjaro are receding rapidly and could soon disappear.

The poverty and agrarian economy that characterize much of the continent make further changes in precipitation a significant concern. More than three-fourths of Africans are directly reliant on rain-fed agriculture. Most of Africa does not possess large-scale water infrastructure. This means that droughts and floods have life-or-death consequences across a wide range of environments. Decreasing precipitation in some areas could be catastrophic to agriculture and livestock and reduce the amount and quality of water available for people to drink. Precipitation changes will also disrupt delicate hydrological balances in much of the continent, impacting a range of natural ecosystems, disrupting water-dependent socioeconomic activities, and making overall water management more difficult. In some areas, hydroelectric production could be impacted (Gabre-Madhin and Haggblade 2004).

Climate impacts will also exacerbate preexisting environmental trends. Many coastal resources that have declined from overfishing, pollution, and coastal erosion now face threats due to ocean warming and acidification. Land degradation—caused by a combination of intensive farming and rapid population increases—is already accelerating as a result of increased drought and changing climatic conditions (Davidson et al. 2003). Increasing demand, contamination, pollution, and precipitation changes are impacting water quality and availability.

Many African leaders and their allies increasingly understand the threats that climate change poses and how economic constraints hamper their countries' ability to adapt. The links between climate and development have received particular attention (Pielke et al. 2007). Indeed, since at least the 2002 World Summit on Sustainable Development, held in Johannesburg, South Africa, African governments, international organizations, and donor countries have searched for ways to use global climate concerns to address the interrelated challenges of African development, climate adaptation, and climate mitigation (Davidson et al. 2003).

For example, many of the plans supported by the National Adaptation Programme of Action (NAPA) developed by African countries as part of the UNFCCC process, include related development activities in the transportation, agriculture, coastal management, and energy production and efficiency sectors. Some countries then build specific NAPA recommendations into their national development and poverty reduction strategies. NAPA recommendations have also been incorporated into the development strategies of donor organizations, including the World Bank and the Global Humanitarian Forum, an international aid organization led by former UN Secretary General Kofi Annan (World Bank 2004; Global Humanitarian Forum 2008).

African countries and their governments vary significantly and possess many different and often conflicting interests on a variety of international, political, economic, and cultural issues. Nevertheless, in international climate negotiations, African governments band together in the Africa Group and Group of 77 (which includes all developing countries) to give their concerns greater political weight. The Africa Group consistently notes that African countries collectively contribute just 5 percent of current global emissions and far less of the overall historical total. It argues that industrialized countries should make deep and long-term emissions reductions before developing countries are called on to do the same. It also supports greater provision of financial and technical aid to assist developing countries in further developing their economies and infrastructures, to assist them in adapting to climate change, and as compensation for climate-related damages. Some member countries of the Africa Group call for programs to establish non-fossil-fuel electricity production in areas ripe for particular technologies, such as solar in the Sahara, geothermal in East Africa, hydro in the wet tropics, and waste-to-energy plants in the largest cities. Such projects would fulfill both development and mitigation goals by preventing future emissions and eliminating some current ones. The resulting economic activity might also provide more local resources for adaptation. African governments also call for more projects via the Clean Development Mechanism (CDM) and support development of financial and other incentives to assist African and other developing countries in reducing deforestation and other land-use practices that contribute to climate change.

Negotiations on all these issues continue, although the short- and perhaps long-term prospects for an agreement on damage payments appear remote. Efforts to develop climate-friendly power facilities are largely taking place outside the UN-FCCC process, with individual African nations working with particular donor countries and international organizations to establish power plants, with mixed initial success. As outlined below, efforts to develop incentives for preventing deforestation and for providing additional resources are proceeding positively. On adaptation funding, in 2007 at the climate negotiations in Bali, governments agreed to finance and initiate operations of an Adaptation Fund to provide developing countries with additional resources. The fund is an acknowledgment that many of the countries that have contributed the least to climate change will still suffer significant consequences and will need assistance to prepare for its effects. The full extent and ultimate success of these programs, if indeed they proceed any further, will be determined in the future.

China

China is the world's most populous country, with more than 1.3 billion people. It has more than 50 cities with 1 million inhabitants, whereas the United States has only 9. China is also the third-largest country in terms of land mass. For several decades, China's economic growth has been among the fastest in the world. China's gross domestic product (GDP) has increased, on average, more than 8 percent annually since 1978. China's economy is now the fourth largest in the world, trailing only the United States, Germany, and Japan (CIA *World Factbook* 2008).

This spectacular economic growth has lifted hundreds of millions of Chinese people out of poverty. From the late 1970s to 2004, China's poverty rate dropped from 65 percent of the population to just under 10 percent (Asian Development Bank 2007). As a result, millions of people have increased their standard of living, gained access to health care, and sent their children to school.

At the same time, more than 100 million people remain in extreme poverty, and far more live below what is considered poverty in the United States and Europe. In addition, the gains associated with China's economic growth have not been evenly distributed, resulting in growing inequality (Asian Development Bank 2007). Consequently, the Chinese government and much of the public (as indicated by what limited independent research that exists) supports continued strong economic growth.

China's tremendous economic growth, utilization of natural resources, dramatic building boom, focus on private wealth creation, and weak or poorly enforced pollution standards have caused significant environmental damage. China's urban air pollution is among the worst in the world. Rivers and lakes near industrial centers are increasingly contaminated. Many factories and a broad range of mining facilities do not adequately manage their operations in an environmentally sound manner and thus seriously contaminate the natural surroundings. These and other environmental issues are the subject of an increasing number of press reports and academic studies in both China and the United States.

Chinese leaders understand this situation. Many new laws and initiatives have been created in the last decade to address the impact of Chinese industrialization, and senior government officials voice public concern that pollution must be controlled in order to protect human health and the environment. Some also note that pollution could become so severe that it may actually harm economic growth by impacting human health, driving away foreign investment, or harming China's global image. In this regard, China took significant steps to limit air pollution during the 2008 Olympic Games in Beijing by, for example, shutting down some factories in the surrounding area for three weeks prior to the start of the Olympics and limiting automobile traffic in the capital city immediately prior to and during the games.

History shows that both the United States and Europe went through similar periods of escalating pollution during their industrialization. Thousands died from air pollution in the worst of London's "killer fogs" in 1952. In the United States, Los Angeles's air pollution was legendary; fish caught in many rivers and lakes in different parts of the country were unsafe to eat; toxic waste contamination forced factories, individuals, and parts of several towns to relocate (Love Canal is the most famous example); and the Cuyahoga River literally caught fire when a spark from a blow torch ignited large amounts of oil and debris floating in the river. Nevertheless, it is the immense scale of the transition in China—its population, economy, economic growth, pollution levels, and possible future pollution—that cause concern for China's impact on the planet as a whole.

This is certainly the case with regard to climate change. China recently became the world's largest greenhouse gas emitter, passing the United States. Equally important, China's emissions are growing at about 11 percent per year, 10 times the U.S. growth rate (Richerzhagen and Scholz 2008). China also relies on coal for up to 70 percent of its energy production, which, in addition to producing air pollution and other environmental impacts, releases vast greenhouse gas emissions. Moreover, given China's immense coal reserves, nearly all observers believe China's use of coal will continue to grow.

As a result, China faces increasing pressure to take action to curb its greenhouse gas emissions. Some industrialized countries stress that it will not be possible to mitigate global warming without greenhouse gas cuts by China. Some governments and business interests argue that as a major economic competitor, China gains unfair advantage if it does not restrict its emissions. These arguments see present China as a "free rider," enjoying the benefits of greenhouse gas cuts by others without taking significant steps to reduce its own contribution to the problem (Zhang 2000; Baumert, Herzog, and Pershing 2005; Richerzhagen and Scholz 2008).

Some governments also express concern about "emissions leakage." They worry that until China starts reducing emissions, efforts within their own countries might be undercut if businesses decide to move emissions-intensive activities to China (Richerzhagen and Scholz 2008). Although little evidence exists that this has occurred to date, it remains a concern, particularly because China already experiences significant amounts of international investment, which drives greenhouse gas emissions.

China will not escape the impacts of climate change (Jiahua 2005; MOST 2007; IPCC WG II 2007), and knowledge of this has produced some internal calls for more climate-friendly energy policies. Some impacts are already being observed. China's National Assessment on Climate Change reports that the mean surface air temperature in the country has increased 0.5 to 0.8°C during the past 100 years, glaciers in the northwest (which supply drinking water) have decreased 21 percent since the 1950s, nearly all major rivers have shrunk, and the frequency and intensity of extreme weather events have changed significantly (MOST 2007).

Projections for the next 20 to 100 years show China experiencing increases both in surface air temperature and in annual precipitation throughout many parts of the country (MOST 2007). By the middle of the century, the annual average temperature in China could rise by as much as 3.3°C, and although parched northern China is expected to experience more precipitation, water shortages will likely increase because of faster evaporation caused by higher temperatures. Drought, heat waves, and other extreme weather are projected to hit China more often, and could impact food production and water availability in certain areas (MOST 2007). China has also publicly recognized the impacts of climate change as a global and regional security issue, particularly with respect to environmental refugees and contested resources with bordering countries (Jiahua 2005).

China has begun taking a number of steps to address greenhouse gas emissions. These likely come in response to concerns about several interrelated issues, including climate change, energy efficiency, air pollution, long-range planning, and international opinion. To oversee this process, the government created the National Coordination Committee on Climate Change. This cross-ministries body develops, deliberates on, and coordinates climate-related policy issues, including both domestic action and international negotiations, although major issues are also submitted to the State Council for direction and guidance.

The most important development to date occurred in July 2007, when China announced the National Climate Change Program. This huge new policy package focuses on five key areas: mitigation, adaptation, technology cooperation, participation in the Kyoto Protocol, and participation in regional efforts. The most important specific policy goals include increasing renewable energy use to 15 percent by 2020; reducing energy intensity by 20 percent by 2010; and cutting major pollutant emissions by 10 percent during the same period (MOST 2007).

These new goals follow a series of measures, now nearly 20 years old, aimed at energy efficiency and conservation. While not designed specifically to lower greenhouse gas emissions, these policies did reduce their rate of growth (Luukkanen and Kaivo-oja 2002). One study estimates that without these measures China's emissions would be almost 50 percent higher than they are today (Zhang 2000).

China has also initiated efforts to reduce air pollution, including introducing regulations on inefficient factories, cement plants, cars, trucks, and especially coal-burning boilers, furnaces, and power plants. If successful, these policies will help pave the way for cleaner, more efficient replacements, improve air quality, and reduce greenhouse gas emissions. China has also engaged in large-scale reforestation and afforestation projects and included climate issues as part of its long-term sustainable development strategy (Jiahua 2005). In 2005, China began large-scale construction of desert solar power plants and rooftop photovoltaic systems connected to the nation's power grid. The use of solar energy is already growing as fast or faster as anywhere else in the world, and it is expected that this growth will accelerate in coming years (Mahabir 2008). China is also installing a large number of wind turbines, and it is commonly believed that China could become the largest wind market in the world within 20 to 30 years. In addition, China's central bank recently developed a tentative outline for a domestic emissions trading scheme that could cover a great variety of effluents, including greenhouse gases.

Interestingly, an increasing number of Chinese government officials, academic experts, and business leaders perceive opportunities for economic gain via these and other mitigation activities. While some measures to curb greenhouse gas emissions could negatively impact short-term economic growth, China stands to gain a great deal from the transition to a more energy-efficient and climate-friendly economy and could become an important actor in the research, development, and manufacture of clean energy infrastructure (Richerzhagen and Scholz 2008). China is already the largest exporter of solar panels in the world (World Bank 2008). Sales of solar water heaters in China are 10 times that in Europe, and its overall solar energy industry expects 20 to 30 percent annual growth (Mahabir 2008). Markets for China's current and future set of solar and wind energy products would benefit from stronger climate policies around the world.

China's position in international climate negotiations is also evolving. China and other developing countries still emphasize the historical responsibility of the industrialized countries in creating the climate change problem. China also draws attention to the immense development challenges it faces and the fact that its per capita emissions are just one-fifth those of the United States. On this basis, China—along with all other developing countries—argues that industrialized countries must initiate serious emissions reductions before China and other developing countries begin such efforts.

Provided the United States and other industrialized countries accept large emissions reductions, China has signaled a new willingness to agree upon some type of mitigation activity. For example, at the conclusion of the 2007 climate negotiations in Bali, developing countries agreed that any new treaty also should include "nationally appropriate mitigation actions by developing country Parties" (Bali Action Plan 2007).

The willingness of China to participate in serious greenhouse gas reductions is crucial to the success of global mitigation efforts. China is already the world's largest greenhouse gas emitter, and these emissions are growing rapidly. At the same time, China's per capita emissions are much less than those of the United States, and hundreds of millions of Chinese people still live in extreme poverty. As a result, China must find a delicate balance that allows it to continue its economic growth while limiting both the climate risk and the diplomatic costs associated with high emissions.

India

India is the world's second-most populous country and the world's largest democracy, with an estimated population of 1.15 billion people. The seventh-largest country in terms of area, India is about one-third the size of the United States but has about three times the population. Almost 70 percent of India's people live in rural areas, although migration to cities has led to a dramatic increase in the country's urban population. India's largest cities are among the world's most populous.

India has tremendous exposure to climate-related impacts, with some occurring already. Monsoons are becoming more severe and less predictable. Glaciers in the Himalayas are melting at unprecedented rates. Both droughts and floods are occurring more frequently. Sea level rise has started to affect coastal mangrove forests, an essential natural defense against storms, in southern India and Bengal (Padma 2006). These and other impacts of climate change are expected to increase significantly.

The IPCC reports that India will experience a 2.7 to 4.3°C increase in average temperatures by 2080. Most of this increase will be felt during the postmonsoon and winter seasons and may lead to a rise in extreme weather events, especially in the northwest, as well as reduced precipitation along India's east coast. Changing weather patterns, especially increased droughts and storms, will significantly impact agriculture, particularly subsistence agriculture that sustains hundreds of millions of people in rural areas. Mountain glaciers and snow packs, which feed India's great rivers and supply water to hundreds of millions, will decline significantly. Sea levels will rise, inundating low-lying agricultural land, increasing erosion, and displacing coastal residents living just above sea level. The combination of rising seas and more extreme storm events will also increase the flood risk for the 25 percent of India's population (nearly 290 million people) that live on the coast. A sea level rise of only 1 meter would displace at least 7 million Indians and submerge 35 percent of neighboring Bangladesh (Parikh and Parikh 2002).

Though India's economy is robust, it is still very poor and will have great difficulty adjusting to climatic changes. India has the second-fastest growing large economy, the twelfth-largest overall economy, the second-largest labor force, and the fourth-largest economy in total purchasing power in the world. In 2006 and 2007, India's GDP grew at an average rate of 8.5 percent (CIA *World Factbook* 2008). While this growth has raised the living standards of millions of people—contributing to improvements in health, education, and basic sanitation across the country—India still suffers from very high levels of extreme poverty, illiteracy, and malnutrition. A quarter of all Indians, nearly 290 million people, live on 40 cents per day (Government of India 2007). Tens of millions more live without what most U.S. residents would consider adequate or reliable supplies of food, freshwater, health care, sanitation, and shelter.

Consequently, economic growth remains a priority for the Indian government. Though it may be justified, this position does pose a challenge for global climate policy. India primarily relies on coal for its energy needs, and its energy consumption is expected to quadruple by the year 2030 (Padma 2006). With greenhouse gas emissions of more than 1.5 billion tons a year, India is already the world's seventh-largest emitter, and its emissions are currently growing at 4.2 percent per year (Sharma, Bhattacharya, and Garg 2006). As India's energy consumption rises, its emissions are expected to increase dramatically, perhaps tripling by 2020.

At the same time, India's per capita emissions are still quite small. While the average American is responsible for 24.3 tons of CO_2 annually, the average Indian produces only 1.6 tons (WRI 2007). Also, India's emissions growth rate of 4.2 percent is lower than that of China (5 percent), Brazil (6 percent), and several other developing countries (Sharma, Bhattacharya, and Garg 2006).

Citing its "right to economic development" and very low per capita emissions, the Indian government has largely declined to implement greenhouse gas controls until developed countries reduce their emissions significantly and provide financial and technical assistance to allow developing countries to employ cleaner energy technologies and implement other measures (Parikh and Parikh 2002). India's former Prime Minister Pradipto Ghosh repeatedly told Western leaders that they must get serious about cutting their own emissions if they want global progress on climate change mitigation. He stressed that India seeks a long-term solution based on per capita emissions, which it sees as the fairest way to account for both the industrialized countries' huge share of total greenhouse gas emissions over the last 200 years and their current energy-intensive consumption patterns (Foster 2007). Current Prime Minister Manmohan Singh pledged that while India will keep its per capita emissions below that of developed nations, it will not accept an international agreement that allows developed nations to emit three to four times the per capita rate of India (Sharma 2008).

India and other developing countries assert that greater investment, financial assistance, and technology transfers are necessary if they are to avoid the development pattern set by industrialized countries. Without environmentally friendly technologies, India argues, it will not be able to reduce emissions and engage in the kind of economic growth necessary to combat extreme poverty (Sharma 2008). The CDM, Adaptation Fund, and GEF projects represent meaningful starting points, but the scale of the problem and the importance of the issues involved will require far greater efforts.

Nevertheless, India has begun taking some mitigation and adaptation measures, many linked to broader development activity. In 2001, Delhi became the first capital city to introduce a large public transport system based on compressed natural gas (which produces far fewer emissions than gasoline or diesel fuel). The government has set a target of meeting 20 to 25 percent of national energy demand from renewable sources (Padma 2006). Afforestation and reforestation efforts are underway, although not on the scale believed needed. And government officials are working to promote energy conservation, energy efficiency, alternative fuels, and renewable energy technologies (Bhandari 2006).

These and other measures will certainly moderate the emissions intensity of India's economic growth, and perhaps reduce the rate of emissions growth, but they will not produce actual reductions in greenhouse gas emissions. Unless the situation changes dramatically, that type of policy development will be dependent on a new global agreement that assists India and other large developing countries in adopting cleaner energy technologies.

Brazil

Brazil is the fifth-largest country in the world, comprising more than 8.5 million square kilometers. Brazil's population of almost 200 million is the world's fifth largest. It maintains the ninth-largest economy, plays a major role in political and economic issues in South America, and is the seventh-largest emitter of greenhouse gases.

Brazil's emissions profile is different from most countries. Less than 20 percent of Brazil's total greenhouse gas emissions result from energy production and use (Chandler et al. 2002). The primary reason for this is Brazil's use of non-fossil-fuel energy sources. Large hydropower stations produce about 80 percent of Brazil's electricity, and Brazil is also the world's largest producer and consumer of ethanol. Sugarcane and wood-based ethanol are responsible for large amounts of national energy consumption. Indeed, Brazil is the world leader in the use of ethanol for transportation.

The bulk of Brazil's greenhouse gas emissions come primarily from deforestation (up to 70 percent of total emissions), other land-use changes, and agriculture (La Rovere and Pereira 2005). Deforestation is driven by the demand for wood and agricultural land, primarily for large soybean plantations, ethanol crops, and extensive cattle rearing. Despite a series of national and international initiatives, Brazil continues to experience rapid deforestation, which releases millions of tons of carbon and methane that was previously embedded in the forests and soils (Belmonte 2007).

Cattle also contribute to Brazil's total greenhouse gas emissions. Cattle release methane through their digestive processes. According to Brazil's Ministry of Science and Technology, in 1995 Brazilian methane emissions totaled about 10 megatons, 80 percent of which came from beef cattle (as opposed to cows raised for milk). Converted to CO_2 equivalent, that equals 233.7 megatons—only slightly lower than the 248.4 megatons of greenhouse gases emitted by the energy sector the previous year (La Rovere and Pereira 2005).

Many regions of Brazil are quite vulnerable to climate change (information in this section is from IPCC WG II 2007; La Rovere and Pereira 2005; Mendes 2007). Anticipated changes in rainfall patterns will strain water resources, especially in the drought-affected northeastern region of the country, and also impact hydropower production. Floods, already a serious problem for several regions, may increase. Rising temperatures will likely expand habitats for several insects that act as vectors for diseases, including mosquitoes, which transmit dengue fever and malaria, and assassin bugs (*Tripanosomiasis americana*), which transmit Chagas disease. Coastal areas, where the bulk of the population and economic activities are concentrated, are vulnerable to rising sea levels, erosion, and storm surges. Coral reefs along Brazilian coastlines could suffer from rising temperatures, ocean acidification, and increased coastal runoff.

Climate change will impact Brazil's tremendous agricultural sector. In some areas, precipitation decline will make Brazil's most important crops (soybeans, corn, and coffee) more difficult to grow. Rising temperatures will also impact crop production in areas where summer temperatures are already quite high.

Brazil's rainforests, home to the greatest concentration of biodiversity in the world, also face serious threats from climate change. Deforestation is already causing changes in regional climate, as increased drying of surface lands affects precipitation patterns and increases the frequency of spontaneous fires in the Amazon region. Rising temperatures from global climate change will exacerbate this situation. The combination of deforestation and global climate change could cause the conversion of large tracts of the rainforest ecosystem to dry grassland.

Some climate impacts are already underway. In 2005, Brazil experienced its worst drought in half a century as well as the highest temperature ever recorded in the country, 44.6°C, or 112°F. Nevertheless, Brazil has refused to adopt national or international emissions-reduction commitments or take long-term measures to stop deforestation. Pointing to its low energy-related emissions and development needs, Brazil argues that developed countries have contributed most to the problem and therefore must take the most significant steps to combat it. Brazilian efforts to combat deforestation seem to gain or lose strength in response to a variety of domestic economic and political considerations. Worries for potential economic losses from reduced timber sales and smaller increases in agricultural production remain the principal concerns.

Brazil has joined other rainforest nations to advocate the creation of new incentive mechanisms within the UNFCCC/ Kyoto umbrella that would grant them carbon credits by reducing or avoiding deforestation. This concept found general agreement during the 2007 Bali climate negotiations, although many technical and methodological issues must be finalized. Avoided deforestation is widely recognized as one of the least expensive ways to reduce greenhouse gas emissions and would also greatly assist efforts to protect biodiversity. A policy that provides carbon credits within the emissions trading or CDM structures of the Kyoto Protocol would also deliver substantial development benefits to countries such as Brazil, create important economic incentives to landowners and the government, and provide an opportunity for developing countries with forest reserves to take on emissions reductions as part of a new global policy structure.

Several other new initiatives have arisen recently as part of the increasing global attention to reduce emissions from deforestation. Norway has pledged to spend up to $500 million to fund rainforest protection projects in developing countries. Industrialized countries have agreed to create a Forest Carbon Partnership Facility, with an initial set of pledges of $165 million, to support activities to address the true drivers of deforestation and support the efforts of rainforest countries to build the capacity to reduce emissions from deforestation. To create an institutional infrastructure to process additional programs, Brazil's state-run development bank established a fund to collect international donations for Amazon preservation initiatives.

Russia

Stretching across 11 time zones, the Russian Federation is the largest country in the world, covering more than one-eighth of the Earth's land mass. It is home to a wide range of ecological diversity, including tundra, grasslands, coniferous forests, mountains, and arid regions. At 141 million, its population is the tenth largest in the world. Driven by steadily increasing prices for its energy exports, the Russian economy is booming. Since rebounding from the financial crisis of 1998, Russia's GDP has grown at an annual average of 7 percent. Russia is the world's largest exporter of natural gas, is the second-largest exporter of oil, and contains huge coal reserves. Although diversifying, the country remains largely dependent on revenue from energy exports and available domestic energy sources (CIA *World Factbook* 2008).

Russia's vast energy endowments are convenient, since the country is also the third-largest energy consumer in the world (CIA *World Factbook* 2008). Political calculations and central planning led the former Soviet Union to build cities in the cold and far-flung northern and western regions, creating a legacy of high heating and transportation costs (Hill and Gaddy 2003). Furthermore, the energy and industrial infrastructure Russia inherited from the Soviet Union was in many cases antiquated and inefficient. Partly as a result, the Russian economy is driven by energy-intensive activities, with energy use per unit of GDP 3.1 times larger than the EU.

Along with other heavy consumers of fossil fuels, Russia is a relatively large emitter of greenhouse gases, although emissions declined significantly due to the severe economic recession that accompanied the breakup of the former Soviet Union. By 1999, emissions totaled just 61.5 percent of those in 1990 (IACRF 2002). In recent years, Russia's greenhouse gas emissions have started to rise quickly. In 2006, for example, they increased 8.5 percent. With Russia's economy and greenhouse gas emissions growing rapidly, the country is critical to the success of any future climate change agreement.

Russia was also critical to the creation of the Kyoto Protocol. In order to come into force, the Kyoto Protocol required ratification by at least 55 nations, accounting for at least 55 percent of developed-country greenhouse gas emissions in 1990. After the United States refused to ratify the protocol, Russia's participation became necessary to achieve this threshold. After internal debate and significant lobbying by the EU, then Russian President Vladimir Putin approved the treaty on November 4, 2004. The protocol entered into force 90 days later. Should a future global agreement require similar thresholds, Russian assent could again prove decisive.

Russia clearly benefits from the terms of the Kyoto Protocol. The protocol requires Russia to reduce its emissions back to 1990 levels by the 2008 to 2012 commitment period. However, because of the precipitous economic decline that followed the collapse of the Soviet Union, Russia's greenhouse gas emissions were already far below 1990 levels by the time it ratified the Kyoto Protocol in 2004. As a result, Russia's baseline, and thus its allotment of tradable permits under the emissions trading provisions, included emissions that no longer existed. This threshold allows Russia to sell these excess emission allowances, sometimes called "hot air" by critics, without taking any steps to reduce actual emissions. In theory, Russia could even package emission credits with exports of fossil fuels, but it continues to face implementation challenges in developing the institutions and systems it needs to benefit fully from the Kyoto mechanisms (Hagem and Maestad 2006). The pace at which Russia is able to meet these challenges will have implications for both Russia and other Kyoto parties that look to the carbon market to meet their own targets (Korppoo and Moe 2007).

Like all countries, Russia will experience negative impacts from climate change. Climate change is already seen as a contributing factor in the increased incidence of floods in Russia and in the receding areas of tundra, which brings with it a host of problems, including ecosystem collapse (Danilov-Danilyan 2007). In the Kuban and Stavropol regions of Russia's breadbasket, droughts are expected to become more common and more severe. Melting permafrost will damage delicate ecosystems and could further exacerbate global warming by releasing accumulated methane and carbon dioxide gas. Melting permafrost will also force Russia to reinforce or even rebuild large amounts of industrial infrastructure—including roads, pipelines, and transmission lines that now sit on frozen ground but that will begin to sink as the ground melts. Rising temperatures will also contribute to the reduction of steppe pine forests in Siberia and the deterioration of coniferous forests in central Russia. As a result, millions of geese, eiders, brants, and other birds will lose their nesting grounds in the next 20 to 40 years (Danilov-Danilyan 2007).

Yet Russia is one of the few countries that could see some benefits from climate change, at least initially. Potential gains include extended growing seasons, reduced ice in Arctic shipping lanes, reduced heating costs, less expensive mining and oil and gas recovery, potential new discoveries of oil and gas in areas of the Arctic Ocean currently covered in ice, and increased economic activity in the far north.

Russia has taken no significant action to restrict domestic greenhouse gas emissions, although achieving former President Putin's target of doubling Russia's GDP in 10 years will require dramatic improvements in energy efficiency that will reduce their growth (Korppoo and Moe 2007). No authoritative reports indicate that Russia is likely to mandate significant domestic emissions cuts or take a leadership role in creating strong global controls. As a former superpower eager to rebuild its industry and economic position in the world, many government priorities appear focused on economic growth, which will likely also mean increasing energy exports, energy use, and greenhouse gas emissions. Russia's oil and gas interests also make it unlikely that Moscow would support global greenhouse gas reduction policies that would drastically reduce demands for its exports (Hagem and Maestad 2006).

Some observers argue that these types of political, economic, and energy interests, rather than environmental ones, will drive Russian climate policy for the foreseeable future. In their view, specific tactical and strategic interests prompted Russia to sign the Kyoto Protocol, including the opportunity to benefit from trading emissions credits and gaining EU support for Russia's hope of entry into the World Trade Organization (WTO). Once these motivations are no longer important, analysts believe Russia will likely choose not to meet environmental targets (Korppoo, Karas, and Grubb 2006).

At the same time, Russia does have some energy- and diplomatic-related interests that may cause it to play a more productive role in climate issues. Russia is involved in a complex and sometimes tense relationship with the EU over natural gas supplies. Russia wants to continue selling natural gas and to secure long-term contracts but has sometimes used the interruption of gas supplies as a political tool in relationships with former Soviet states or allies. Governments in some Western and Central European countries want this practice to stop and also want to reduce their oil and gas dependence on Russia. Finding the correct long-term political and economic mix to satisfy all parties could be tricky (Rachman 2007). Similarly, Russia is engaged in what will likely be a long-term diplomatic competition with Canada, Denmark, Norway, and the United States for recognition of sovereignty over particular Arctic shipping lines and sites for oil and gas exploration that are emerging as sea ice recedes. Russia's invasion of Georgia also damaged Russia's relationship with the United States. While security issues still outweigh environmental issues in most countries' strategic thinking, Russia could choose to show flexibility on climate issues in the future as part of an effort to improve relations. Russia may find that supporting a new climate treaty could be helpful for getting other countries to show flexibility on energy issues or other matters, including diplomatic support for full entry into the WTO, which Moscow still needs.

Perhaps as a result of such competing interests, Russia has kept its position on global climate policy relatively opaque. Yet despite—or perhaps because of—its unwillingness to take a bold stance with regard to international policy, Russia continues to be an important player in the international climate policy.

European Union

The European Union is an economic and political confederation of 27 developed countries located primarily, but not entirely, in Europe. The EU stretches across 4 million square kilometers and has nearly 500 million citizens. It generates roughly one-third of the world economy and about 15 percent of global greenhouse gas emissions, or 4.5 billion tons of carbon emissions annually.

The EU maintains a single, integrated economic market that allows for the free movement of capital, goods, services, and people across national borders. Member states develop and observe standardized regulations on many aspects of their economic, agricultural, energy, environmental, trade, and international policies. Fifteen nation-states have adopted the euro as their common currency. The EU governing structure includes the European Council (heads of states), the presidency (which rotates among the EU member states every six months), the elected European Parliament, the Court of Justice, and the European Commission. The European Commission is somewhat equivalent to the executive branch in the United States; it is responsible for implementing regulations, developing and proposing new policies, and running the day-to-day affairs of the European Union. During international negotiations, the EU speaks for all the member states, giving its positions special importance.

EU countries have already begun experiencing climate-related impacts, and far more serious events will occur in the future. The continent warmed by an average 0.8°C during the last century (IPCC WG II 2007). A historic heat wave in 2003 contributed to the deaths of 50,000 people (EEA 2004). Snow cover has decreased and glaciers are receding in all parts of Europe, impacting winter tourism and threatening the viability of some ski resorts. Precipitation patterns have begun to change. Flood risks are projected to increase for most of Europe, and the continent will likely see more frequent and more damaging storms—such as those that fueled serious flooding in six European countries in 2005. In the already-arid Mediterranean regions, more frequent drought will present a threat to agriculture, though some of these impacts may be balanced by warmer temperatures that could extend the growing season in Northern Europe (Kerr 2002). Habitats will shift, some insects will expand their range, and sea level rise will impact coastal regions.

The EU has taken a strong leadership position with regard to climate change. As early as 1990, the energy and environment ministers of the EU's then 15 members agreed that emissions should be no higher in 2000 than they were in 1990 (Gummer and Moreland 2000). In 2002, the EU, along with its individual member states, ratified the Kyoto Protocol and agreed to reduce EU-wide emissions by 8 percent below 1990 levels by 2008–2012.

To achieve this target, member states apportioned reductions among themselves. Wealthier countries generally took on larger cuts. Targets ranged from –28 percent for Luxembourg (which has the highest per capita GDP in the world) to +27 percent for Portugal (with the lowest GDP in Western Europe) (Michaelowa 2004). Other targets include Germany at –21 percent, the United Kingdom at –12.5 percent, and France (which already has relatively low emissions because of nuclear power) at +/– 0 percent. As the EU has grown, all new states (with the exception of Cyprus and Malta) have accepted reduction targets of 6 to 8 percent below their base year (EEA 2007).

EU states can choose to undertake reductions through domestic policies or the flexibility mechanisms under the Kyoto Protocol. In general, domestic policies are shaped by policy directives issued by the European Climate Change Program (ECCP). Some of the ECCP priority areas for domestic regulation include renewable energy, efficiency standards, energy tax reform, and methane capture (EEA 2007).

As outlined in Chapter 1, the three Kyoto flexibility mechanisms are the CDM, Joint Implementation (JI), and emissions trading. The CDM allows developed countries to get credit for investing in emissions reductions in developing countries. JI gives credit for projects in other developed countries. By the end of 2006, EU countries had purchased more than 540 million tons worth of credits from CDM and JI projects for the 2008–2012 commitment period, and this figure is growing (EEA 2007).

In the third flexibility mechanism, emissions trading, governments set a cap on total emissions and then distribute emissions rights to regulated entities. A company that reduces its emissions by more than the required amount is allowed to sell its excess credits to those unable or unwilling to achieve the required reductions.

Following much discussion of possible international trading systems, the EU took the lead and launched the European Union Emissions Trading Scheme (EU-ETS) in 2005. Large stationary sources accounting for 40 percent of total EU emissions are regulated under the EU-ETS (EEA 2007). This regulation includes power and heat generation, cement and ferrous metals, lime, glass, pulp, and paper companies. Many other sources will be added.

During the initial 2005–2006 trading period, the EU-ETS experienced several problems, the most significant being that governments had inadvertently allocated more allowances than needed when establishing the system. As a result, the supply of allowances exceeded demand and only a small market existed for allowances, lowering incentives to reduce emissions. To address this problem, the European Commission reduced the total number of allowances by 10.5 percent during the 2008–2012 trading period. As a result, it is now estimated that EU-ETS will contribute 150 megatons of CO_2 emissions reductions by 2012, equal to roughly 3.4 percent of EU-15 base-year emissions (EEA 2007).

In 2007, the EU also announced its ambitious new 20 20 by 2020 climate policy. The policy includes unilateral commitments to cut EU greenhouse gas emissions by 20 percent below 1990 levels by 2020. The EU will also increase the share of energy that comes from renewable, non-greenhouse-gas-emitting sources to 20 percent and increase energy efficiency by 20 percent, both by 2020. Further, the EU announced that it would increase its commitment to reduce its greenhouse gas emissions to 30 percent by 2020 if other developed countries (including the United States) make significant commitments as well (EC 2008 provides official details of the *20 20 by 2020* policy and the reasons it was adopted).

The EU announced that the reasons for these actions included the details of the 2007 IPCC and the 2006 Stern reports and the EU's general support for efforts to keep global temperatures from rising more than 2°C above preindustrial levels. Many scientists and policy makers see 2°C as a so-called tipping point beyond which many very serious and potentially irreversible impacts will occur (EC 2007a, 2007b; UCS 2007). Achieving this goal, which the EU has highlighted for nearly a decade, will require industrialized nations to make deep near-term emissions cuts, perhaps to 30 percent below 1990 levels by 2020, and then take the lead in helping reduce global emissions by 50 percent or more from 1990 levels by 2050 (EC 2007b). The EU commitment, which goes far beyond what any other country has announced, demonstrates both the seriousness with which it regards the issue and the promise that such cuts are actually possible.

Organization of the Petroleum Exporting Countries

Created in 1960, the Organization of the Petroleum Exporting Countries (OPEC) is an intergovernmental organization whose members work together to control oil production. Its goal is to keep prices at levels that ensure sufficiently large profits for member states without crippling the economies of OPEC's customers. OPEC's current members are Algeria, Angola, Ecuador, Indonesia, Iran, Iraq, Kuwait, Libya, Nigeria, Qatar, Saudi Arabia, the United Arab Emirates, and Venezuela (OPEC 2008). While several large oil producers are not members of OPEC—

including Canada, Norway, Russia, and the United States—OPEC member countries do contain about 76 percent of proven global oil reserves (World Energy Council 2007).

Climate change will impact OPEC countries negatively, but OPEC's primary concern is how climate mitigation policies could reduce global demand for oil. Accordingly, OPEC and many individual member states have consistently worked against the creation of strong action on climate change.

In the 1990s, for example, OPEC launched a public relations campaign that emphasized real and perceived scientific uncertainties to argue that scientific knowledge of the issue did not justify mitigation measures. Several OPEC members, Saudi Arabia being the most active, still sometimes attempt to slow the global negotiations by arguing over rules of procedure, shifting discussions to procedural issues, making lengthy interventions, and moving discussion to mitigation options unrelated to fossil fuels, among other strategies (Aarts and Janssen 2002; personal observations by the authors). Some OPEC states have worked quietly with the United States to oppose efforts to expand quantified greenhouse gas emission targets, emphasizing the view that reducing emissions from fossil fuels carries high economic costs. OPEC members strongly support China and India's position that they not be required to reduce their growing emissions or scale back their expanding efforts to develop massive auto industries. OPEC also cooperates with business lobbyists representing mainly coal and oil industries in their goal to forestall serious mitigation efforts.

In recent years, and particularly after the 2007 IPCC report, OPEC has toned down arguments involving scientific uncertainty and concentrated more on arguing that oil exporters should be compensated for any loss incurred as the result of global climate policy. It also argues that oil exporters should receive "demand security" guaranteeing a certain range of future oil exports so that the export countries can make long-term decisions about oil investments and production levels (Aarts and Janssen 2002).

The prospects for such guarantees appear slim. Few nations are likely to agree to pay for oil they do not buy, particularly given the windfall profits oil exporters have received in recent years. More practically, the amount of revenue lost to mitigation policies would be impossible to define without knowledge of how the world oil market would have operated without them

(Kassler and Paterson 1997). In addition, given the rapid development of countries like India and China, the International Energy Agency considers it unlikely that world oil consumption will decline in the next 20 years, even if strong global mitigation policies emerge (IEA 2006). Since OPEC must be aware of all these issues, it seems probable that its insistence on discussing compensation is part of its broader strategy of seeking to delay agreement on deep, long-term emissions cuts.

At the same time, OPEC nations do appear more willing to discuss and even act on some aspects of climate change mitigation. In November 2007, OPEC issued the Riyadh Declaration, a statement that indicated its interest in finding a compromise between expected growth in fossil fuel use and the need to address climate change concerns (OPEC 2007). The governments of OPEC's key members have also pledged $750 million to fund studies of energy, its environmental impact, and climate change.

If this movement does represent a change, it may also be tied to growing realizations concerning the negative impacts that climate change will impose on OPEC nations. OPEC countries in the Middle East, as well as Angola, can expect increased temperatures and reduced precipitation. These outcomes could disrupt the development of alternate sources of economic growth, including agriculture and tourism. More extreme storms would impact oil production and transport, as we saw in the Gulf of Mexico in the aftermath of Hurricane Katrina. Increased pressures on health, habitat, food security, and the availability of water could exacerbate preexisting social problems and lead to unrest (Halden 2007).

Summary and Conclusion

Climate change is a global problem. Its impacts on specific countries around the world will vary in response to a host of geographic, climatic, economic, political, and social factors, but on the whole they are expected to be negative—and in some cases, potentially disastrous. Some fragile ecosystems, coastal areas, and low-lying islands will be destroyed. Certain species will not be able to adapt and will go extinct. Many of the world's poorest people will face more hunger, less access to freshwater, more exposure to extreme weather, and more threats from a variety of diseases.

A key factor in determining how soon and how badly human populations will be affected is the ability of a given country to marshal the resources necessary to adapt. Most at risk are the people and natural environments on small island developing states, in Africa, and in India and neighboring countries. Already living in precarious positions, these individuals face rising temperatures, changing weather patterns, increased storms and disease, and rising sea levels that will endanger the very existence of many people, plant and animal species, ecosystems, and even entire nations.

Great variations exist in how key countries are responding to climate change. The sources for these differences are complex but include levels of development, sources of greenhouse gas emissions, national energy policies, key economic sectors, and the expected impacts of climate change. To date, only the EU among the major emitters has committed to making major reductions in greenhouse gas emissions beyond what is called for in the Kyoto Protocol. The United States refuses to participate in any global agreement unless the large, rapidly developing countries take action, a position supported quietly by other countries with regard to future agreements.

Developing countries argue that their low per capita energy emissions and development needs justify delaying meaningful contributions to climate change mitigation until the developed world agrees to take on commitments for deep emissions cuts; until the flexibility mechanisms, market structures, and provisions for financial and technical assistance are improved sufficiently to provide the correct incentives; and until the entire package is judged by developing countries as both effective and fair. This coalition includes China, India, and Brazil, which themselves have very large national emissions. Africa's development and climate challenges are immense, so it seeks major investment and assistance in both areas. AOSIS countries work to remind the world that climate change threatens their very survival. OPEC countries largely work against deep emissions cuts that would imperil oil exports. These opposing positions have so far prevented agreement on long-term global policy after the Kyoto Protocol controls expire in 2013.

Climate change is a long-term global problem that requires a long-term global solution. The arithmetic inherent in the issue makes it clear that, eventually, all countries with significant emissions will need to make deep reductions and all countries with major forests will need to protect and replenish them. The presence of new technologies and innovative policies could allow long-term economic benefits to accompany sufficiently well-crafted national and international policies. This chapter outlined the impacts many key countries are facing as well as their core positions on long-term global policy. What it cannot do is say with certainty if these positions can be brought together in time.

References

Aarts, P., and D. Janssen. 2002. "Shades of Opinion: The Oil Exporting Countries and International Climate Politics." *Review of International Affairs* 3 (2): 332–351.

Adger, W. N., S. Huq, K. Brown, D. Conway, and M. Hulme. 2003. "Adaptation to Climate Change in the Developing World." *Progress in Development Studies* 3 (3): 179–195.

Africa Policy E-Journal. 2001. "Africa: Climate Change." [Online information; retrieved 7/15/08.] www.africaaction.org/docs01/clim0102.htm.

Asian Development Bank. 2007. "Reducing Inequalities in China Requires Inclusive Growth." [Online article; retrieved 7/15/08.] http://www.adb.org/Media/Articles/2007/12084-chinese-economics-growths/default.asp.

Bali Action Plan. 2007. "Report of the Conference of the Parties on Its Thirteenth Session, Held in Bali from 3 to 15 December 2007. Addendum. Part Two: Action Taken by the Conference of the Parties at Its Thirteenth Session." UN Document FCCC/ CP/2007/6/Add.1*. [Online document; retrieved 7/15/08.] http://unfccc.int/resource/docs/2007/cop13/eng/06a01.pdf#page=3.

Baumert, K., T. Herzog, and J. Pershing. 2005. *Navigating the Numbers: GHGs and International Climate Change Agreements.* Washington, DC: World Resources Institute.

Belmonte, R. V. 2007. "Brazil: Amazon Logging Means Short-Lived Prosperity." [Online article; retrieved 7/16/08.] http://ipsnews.net/news.asp?idnews=39328.

Bhandari, P. 2006. "India's Pragmatic Approach to Climate Change." [Online commentary; retrieved 7/16/08.] http://www.scidev.net/en/opinions/indias-pragmatic-approach-to-climate-change.html.

Caribbean Community (CARICOM). 2008. "Caribbean Planning for Adaptation to Climate Change (CPACC) Project." [Online article; retrieved 7/16/08.] www.caricom.org/jsp/projects/macc%20project/cpacc.jsp.

Central Intelligence Agency (CIA) *World Factbook.* 2008. [Online publication; retrieved 7/16/08.] https://www.cia.gov/library/publications/the-world-factbook/index.html.

Chandler, W., R. Schaeffer, Z. Dali, P. R. Shukla, F. Tudela, O. Davidson, and S. Alpan-Atamer. 2002. "Climate Change Mitigation in the Developing Countries: Brazil, China, India, Mexico, South Africa, and Turkey." Arlington, VA: Pew Center on Global Climate Change.

Chen, A. 2006. "The Threat of Dengue Fever in the Caribbean: Impacts and Adaptation." Final Report to Assessments of Impacts and Adaptations to Climate Change, Project No. SIS 06. Washington, DC: International START Secretariat.

Danilov-Danilyan, V. 2007. "Danger of Climate Change Equal to Nuclear War." *RIA Novosti,* June 29. [Online article; retrieved 10/23/08.] http://en.rian.ru/analysis/20070627/67914064.html.

Davidson, O., K. Halsnaes, S. Huq, M. Kok, B. Metz, Y. Sokona, and J. Verhagen. 2003. "The Development and Climate Nexus: The Case of Sub-Saharan Africa." *Climate Policy* 3 (Suppl. 1): S97–S113.

European Commission (EC) 2007a. "Limiting Global Climate Change to 2 Degrees Celsius." [Online press release; retrieved 7/16/08.] http://europa.eu/rapid/pressReleasesAction.do?reference=MEMO/07/16.

European Commission (EC) 2007b. "Limiting Global Climate Change to 2 Degrees Celsius: The Way Ahead for 2020 and Beyond." [Online communication; retrieved 8/20/08.] http://eur-lex.europa.eu/LexUriServ/LexUriServ.do?uri=CELEX:52007DC0002:EN:NOT.

European Commission (EC) 2008. "20 20 by 2020: Europe's Climate Change Opportunity." [Online communication; retrieved 7/16/08.] http://ec.europa.eu/commission_barroso/president/pdf/COM2008_030_en.pdf.

European Environment Agency (EEA). 2004. "Impacts of Europe's Changing Climate: An Indicator-based Assessment." [Online report; retrieved 7/16/08.] www.defra.gov.uk/farm/environment/climate-change/eu/2-%20parry.pdf.

European Environment Agency (EEA). 2007. "Greenhouse Gas Emissions Trends and Projections in 2007: Tracking Progress toward Kyoto Targets." Luxembourg: Office for Official Publications of the European Communities.

Foster, P. 2007. "India Snubs West on Climate Change." *The Telegraph,* December 6. [Online information; retrieved 10/23/08.] www.telegraph.co.uk/earth/main.jhtml?xml=/earth/2007/06/12/eaindia12.xml.

Gabre-Madhin, E. Z., and S. Haggblade. 2004. "Successes in African Agriculture." *World Development* 32 (5): 745–766.

Global Humanitarian Forum. 2008. Home page. [Online information; retrieved 7/16/08.] www.ghf-geneva.org.

Government of India, Planning Commission. 2007. "Poverty Estimates for 2004–05." [Online information; retrieved 7/16/08.] www.planning-commission.gov.in/news/prmar07.pdf.

Gummer, J., and R. Moreland. 2000. "European Union and Global Climate Change: A Review of Five National Programs." Arlington, VA: Pew Center on Global Climate Change.

Hagem, C., and O. Maestad. 2006. "Russian Exports of Emission Permits under the Kyoto Protocol: The Interplay with Non-Competitive Fuel Markets." *Resource and Energy Economics* 28 (1): 54–73.

Halden, P. 2007. "The Geopolitics of Climate Change." Stockholm, Sweden: *FOI*, the Swedish Defence Research Agency.

Hill, F., and C. G. Gaddy. 2003. *Siberian Curse: How Communist Planners Left Russia Out in the Cold.* Washington, DC: Brookings Institution.

Intergovernmental Panel on Climate Change Working Group II (IPCC WG II). 2007. *Climate Change 2007: Impacts, Adaptation and Vulnerability. Contribution of Working Group II to the Fourth Assessment Report of the Intergovernmental Panel on Climate Change,* edited by M. L. Parry, O. F. Canziani, J. P. Palutikof, P. J. van der Linden, and C. E. Hanson. Cambridge, UK: Cambridge University Press.

International Energy Agency (IEA). *World Energy Outlook 2006.* [Online publication; retrieved 7/16/08.] www.worldenergyoutlook.org/2006.asp.

Intra-Agency Commission of the Russian Federation (IACRF). 2002. *Third National Communication of the Russian Federation.* [Online report; retrieved 7/16/08.] http://unfccc.int/resource/docs/natc/rusnce3.pdf.

Jiahua, P. 2005. "China and Climate Change: The Role of the Energy Sector." [Online article; retrieved 7/16/08.] http://scidev.net/en/science-communication/climate-change-in-china/policy-briefs/china-and-climate-change-the-role-of-the-energy-se.html.

Kassler, P., and M. Paterson. 1997. *Energy Exporters and Climate Change.* Washington, DC: Brookings Institution.

Kelly, P. M., and W. N. Adger. 2000. "Theory and Practice in Assessing Vulnerability to Climate Change and Facilitating Adaptation." *Climatic Change* 47:325–352.

Kerr, R. 2002. "Mild European Winters Mostly Hot Air, Not Gulf Stream." *Science* 297 (5590): 2202.

Korppoo, A., and A. Moe. 2007. "Russian Climate Politics: Light at the End of the Tunnel?" Climate Strategies Briefing Paper, April. Cambridge, UK: Climate Strategies.

Korppoo, A., J. Karas, and M. Grubb. 2006. *Russia and the Kyoto Protocol.* London: Chatham House.

La Rovere, E. L., and A. S. Pereira. 2005. "Brazil & Climate Change: A Country Profile." [Online article; retrieved 7/16/08.] http://www.scidev .net/en/policy-briefs/brazil-climate-change-a-country-profile.html.

Luukkanen, J., and J. Kaivo-oja. 2002. "Meaningful Participation in Global Climate Policy? Comparative Analysis of the Energy and CO_2 Efficiency Dynamics of Key Developing Countries." *Global Environmental Change* 12:117–126.

Magrath, A., and A. Simms. 2006. *Up in Smoke 2: Africa and Climate Change.* London: Oxfam.

Mahabir, R. 2008. "China Solar Energy Industry Research and Forecast, 2008–2010." [Online article; retrieved 7/16/08.] http://asiaclean-tech.wordpress.com/2008/01/17/china-solar-energy-industry-research-and-forecast-2008-2010/.

Makan, A. 2007. "Climate Change Threatens Human Rights—Small Island States." [Online article; retrieved 7/16/08.] http://www.alertnet .org/thenews/newsdesk/SP233436.htm.

Mathur, A. 2001. "Adaptation: Challenges and Opportunities in the Caribbean Region." Presentation to the GEF Consultations on NAPA Guidelines, Arusha, Tanzania, February 28.

Mendes, H. 2007. "Brazil Faces Forecast of Heat and Dust." [Online article; retrieved 7/16/08.] http://www.scidev.net/en/features/brazil-faces-forecast-of-heat-and-dust.html.

Michaelowa, Axel. 2004. "Can the EU Provide Credible Leadership for Climate Policy beyond 2012?" In *Kyoto Protocol, Beyond 2012,* edited by Agus P. Sari. Djakarta, Indonesia: Pelangi.

Ministry of Science and Technology of the People's Republic of China. (MOST). 2007. *China's National Assessment Report on Climate Change.* Beijing: China Science Press.

O'Brien, K. L., and R. M. Leichenko. 2003. "Winners and Losers in the Context of Global Change." *Annals of the Association of American Geographers* 93 (1): 89–103.

Organization of the Petroleum Exporting Countries (OPEC). 2007. "The Riyadh Declaration: The Third Summit of Heads of State and Government of OPEC Member Countries." [Online information; retrieved 7/16/08.] http://www.opec.org/aboutus/III%20OPEC%20Summit%20Declaration.pdf.

Organization of the Petroleum Exporting Countries (OPEC). 2008. "Brief History." [Online information; retrieved 7/16/08.] www.opec.org/aboutus/.

Padma, T. V. 2006. "Development versus Climate Change in India." [Online article; retrieved 7/16/08.] http://www.scidev.net/en/features/development-versus-climate-change-in-india.html.

Parikh, J. K., and K. Parikh. 2002. "Climate Change: India's Perceptions, Positions, Policies, and Possibilities." Paris: Organisation for Economic Co-operation and Development.

Patz, J. A., D. Campbell-Lendrum, T. Holloway, and J. A. Foley. 2005. "Impact of Regional Climate Change on Human Health." *Nature* 438: 310–317.

Pielke, R., Jr., G. Prins, S. Rayner, and D. Sarewitz. 2007. "Climate Change 2007: Lifting the Taboo on Adaptation." *Nature* 445:597–598.

Price, T. 2002. "The Canary is Drowning: Tiny Tuvalu Fights Back against Climate Change." *Global Policy Forum.* [Online information; retrieved 10/23/08.] www.globalpolicy.org/nations/micro/2002/1203canary.htm.

Rachman, G. 2007. "The Paradoxical Politics of Energy." [Online article; retrieved 7/16/08.] http://www.economist.com/theworldin/international/displayStory.cfm?story_id=10120070&d=2008.

Richerzhagen, C., and I. Scholz. 2008. "China's Capacities for Mitigating Climate Change." *World Development* 36 (2): 308–324.

Sharma, S. 2008. "India Seeks Climate Technology-Transfer Regime." *Bloomberg News* February 7. [Online information; retrieved 10/23/08.] www.bloomberg.com/apps/news?pid=20601091&sid=a1fCM_8StWDU&refer=india.

Sharma, S., S. Bhattacharya, and A. Garg. 2006. "Greenhouse Gas Emissions from India: A Perspective." *Current Science* 90 (3): 326–333.

Tompkins, E. L. 2005. "Planning for Climate Change in Small Islands: Insights from National Hurricane Preparedness in the Cayman Islands." *Global Environmental Change* 15:139–149.

Union of Concerned Scientists (UCS) 2007. "Leading Climate Scientists Call for Limiting Global Warming to Less than 2 Degrees Celsius above Pre-Industrial Levels." [Online press release; retrieved 7/16/08.] http://

www.ucsusa.org/news/press_release/leading-climate-scientists-0084.html.

United Nations (UN). 2007. "List of Least Developed Countries." [Online information; retrieved 8/20/08.] http://www.un.org/special-rep/ohrlls/ldc/list.htm.

World Bank. 2004. *African Development Indicators.* Washington, DC: World Bank.

World Bank. 2008. "In Search of Clean Energy to Meet China's Needs." [Online article; retrieved 7/16/08.] http://web.worldbank.org/WBSITE/EXTERNAL/NEWS/0,,contentMDK:21589744~pagePK:34370~piPK:34424~theSitePK:4607,00.html.

World Energy Council. 2007. *Survey of Energy Resources, 2007.* [Online publication; retrieved 7/16/08.] http://www.worldenergy.org/documents/ser2007_final_online_version_1.pdf.

World Food Programme (WFP). 2008. Home page. [Online information; retrieved 7/16/08.] www.wfp.org/english/.

World Resources Institute (WRI). 2007. "Climate Analysis Indicators Tool." [Online information; retrieved 7/16/08.] http://cait.wri.org/.

Zhang, Z. X. 2000. "Can China Afford to Commit Itself to an Emissions Cap? An Economic and Political Analysis." *Energy Economics* 22:587–614.

4

Chronology

This chapter presents a brief chronology of significant scientific, policy, and economic developments relating to global climate change. As with any chronology, the events included here represent an indicative list of key developments rather than an exhaustive one. In general, when multiple events are listed for a particular year, they are listed in the order in which they occurred.

Prior to 1750 Carbon dioxide (CO_2) concentrations in the atmosphere are approximately 270 to 290 parts per million (ppm).

Circa 1750 The Industrial Revolution begins in Great Britain in the mid-1700s, when a series of social developments and technological innovations create the conditions for the rapid mechanization of labor, a shift from agricultural to industrial economies, and unprecedented population growth.

1827 Jean-Baptiste Joseph Fourier publishes a paper theorizing that atmospheric gases play a role in warming temperatures at the Earth's surface by preventing heat emitted by the Earth from leaving the atmosphere. He does not understand precisely how this occurs, but the idea is a conceptual leap forward that contributes a great deal to later research.

1850s Thomas Edison begins manufacturing direct current (DC) electric generators.

1859   John Tyndall reports that his experiments indicate that water vapor and carbon dioxide are strong absorbers of radiant heat. He postulates that the absorption of heat by these gases has an effect on climate.

1867   Nikolas August Otto patents the four-stroke internal combustion engine, spawning the rapid mechanization of production.

1870–1910 The second Industrial Revolution occurs as technological advances achieved earlier in Great Britain spread to other countries and scientific discoveries are applied to developing industries. The steel and petroleum industries arise, electricity becomes widely available in Europe and the United States, and the scale of industry grows rapidly in response to developments in mass production and transportation technologies.

1882   Thomas Edison develops the first practical coal-fired electricity generating station in New York City, supplying electricity for household lights.

1890   Nikola Tesla invents the alternating current (AC) generator. This invention, along with the introduction of transformers (which increase voltage as electricity leaves the generation plant and decrease voltage when it reaches its destination), allow electricity to be transmitted longer distances.

1896   Svante Arrhenius publishes an article suggesting that atmospheric temperatures will rise 3°C if the atmospheric concentration of CO_2 doubles.

1897   T. C. Chamberlin produces a global carbon model that incorporates the small fraction of the Earth's carbon stored in the atmosphere. He hypothesizes that ice ages might oscillate with CO_2.

1900   The concentration of CO_2 in the atmosphere reaches 295 ppm.

1920s   The opening of the Persian Gulf and Texas oil fields inaugurates the era of cheap oil.

1925   C. E. P. Brooks suggests that small changes to the Earth's climate may be amplified by self-reinforcing, or positive, feedback cycles. A positive feedback cycle occurs when a system responds to a perturbation by accelerating in the direction of the change.

1930s–
1940s   Milutin Milankovitch develops a climate theory based on seasonal and latitudinal variations of solar radiation received due to changes in the Earth's orbit. The three orbital motions he describes (eccentricity, obliquity, and precession) become known as the Milankovitch cycles.

1938   G. S. Callendar revisits Arrhenius's 1896 publication and argues that increases in CO_2 concentrations could explain recent warming trends.

1945   The United Nations (UN) is established.

1947   Harold Urey discovers that in fossilized materials the presence of different forms of oxygen (O^{18} and O^{16}) varies with the temperature at which the fossil was formed. This discovery offers the potential for constructing a record of past temperatures by analyzing the relative amounts of each form of oxygen in a given material.

          The World Meteorological Organization (WMO) is established.

1950   The concentration of CO_2 in the atmosphere is roughly 310 to 315 ppm.

1951   The first demonstration of nuclear-generated electricity takes place at the Experimental Breeder Reactor site in Idaho.

1952   A toxic mix of dense fog and sooty black coal smoke suffocates 4,000 people in the worst of London's "killer fogs."

1956   Maurice Ewing and William Donner develop a model that suggests that feedbacks involving Arctic ice

1956   cover could promote global climate changes on a sur-
(cont.)   prisingly rapid scale.

         Norman Phillips produces the first general circulation model (GCM), significantly improving upon previous attempts to accurately reflect the observed behavior of the atmosphere at a global scale. Models that build off Phillips's work eventually allow scientists to predict how the Earth's climate will react to changes in atmospheric composition, including rising levels of CO_2.

         Gilbert Plass, an expert in infrared absorption in the atmosphere, calculates that adding CO_2 to the atmosphere will have a significant effect on the radiation balance and that a doubling of CO_2 from preindustrial levels could lead to an increase of 3.6°C in global average temperature.

1957–1958 The International Geophysical Year draws attention to climate research and provides funding for further work.

1957   Roger R. Revelle finds that the ability of the ocean to absorb CO_2 has been overestimated in previous studies.

         Revelle and Hans Seuss raise concerns that the high level of CO_2 emissions from industrialization might alter the composition of the atmosphere.

1958   Charles David Keeling begins the first constant monitoring of global CO_2 levels at the Mauna Loa Observatory in Hawaii.

1960   Keeling publishes the first paper describing an increase in atmospheric CO_2 levels based on records from Mauna Loa.

1961   The U.S. Department of Meteorology reports that global temperatures rose until about 1940 but declined since that time. The report acknowledges that increased CO_2 should lead to increasing temperatures but cannot offer a definitive explanation for the recent cooling trend.

In a speech given at the UN, U.S. president John F. Kennedy urges global cooperation in weather science and prediction. The UN subsequently adopts several resolutions encouraging collaboration between the International Council of Scientific Unions (ICSU) and the WMO.

1963   Building on work by Gilbert Plass, Fritz Möller produces a one-dimensional climate model, which simulates the effect of solar radiation and greenhouse gases in a vertical column of atmosphere. In Möller's model, a doubling of CO_2 creates high levels of atmospheric water vapor and unstable positive feedback cycles. His work stimulates development of more complex computer models.

1965   The first major conference to address climate, Causes of Climate Change, convenes in Boulder, Colorado. Though a wide range of academic disciplines is represented, the conference fails to attract significant political or media attention.

1967   The WMO and ICSU establish the Global Atmospheric Research Program to enhance climate science and short-range weather prediction.

Ed Lorenz publishes work that shows the climate system to be chaotic and inherently difficult to predict.

1968   John Mercer, a glaciologist at the Ohio State University, draws attention to the West Antarctic Ice Sheet by suggesting that its possible collapse would raise sea levels dramatically.

The Massachusetts Institute of Technology (MIT) hosts a month-long scientific symposium at which 68 of the world's leading scientists work to achieve consensus on the principal threats to the environment,

1968   identifying "pollution-induced changes in climate"
(cont.)   among them. The results of the conference are pub-
              lished as the "Study of Critical Environmental Prob-
              lems" and presented at the 1972 Stockholm Conference
              on the Environment.

              Reid Bryson presents evidence that man-made pollu-
              tion in the form of aerosols is causing global cooling.

1971   MIT hosts a three-week conference titled A Study of
              Man's Impact on Climate, at which the world's lead-
              ing scientists try to achieve consensus on threats to
              the climate system. The results are published in the
              volume *Inadvertent Climate Modification*, which reports
              the danger of a rapid and serious global change
              caused by humans and calls for an organized research
              effort.

1972   The United Nations Conference on the Human Envi-
              ronment convenes in Stockholm. The conference, the
              first major global governmental meeting focused on
              the environment, issues a declaration of common
              principles to guide the preservation of the human en-
              vironment. The meeting also results in creation of the
              United Nations Environment Programme (UNEP).

1973   The Organization of the Petroleum Exporting Coun-
              tries (OPEC) imposes an oil embargo on the United
              States and the Netherlands because of their support
              for Israel in the 1973 Arab-Israeli War. Short-term oil
              and gasoline shortages and price increases result.
              When the embargo is lifted after six months, crude oil
              prices have tripled. The event underscores the United
              States' vulnerability to oil-supply disruptions and
              generates increased interest in alternative energy
              sources.

1974   Sherwood Rowland and Mario Molina publish their
              discovery that chlorofluorocarbons (CFCs), widely
              used as refrigerants, solvents, aerosol-product propel-
              lants, and foam-blowing agents, threaten the ozone
              layer.

1975   Syukuro Manabe's and Richard Wetherald's simpli-
            fied three-dimensional GCM adds to the growing
            number of models dedicated to assessing the effect of
            greenhouse gases on the Earth's climate. Their model
            shows that an increase in greenhouse gases will result
            in the warming of the Earth's surface and a more in-
            tense hydrological cycle.

            Veerabhadran Ramanathan finds that CFCs are also
            strong greenhouse gases.

1976   Sediment cores taken from beneath the ocean floor re-
            veal a relationship between Milankovitch cycles and
            climate. This relationship emphasizes that the ice-
            albedo feedback mechanism is a key component to
            monitoring the climate system.

1977   Cesare Marchetti is the first to propose the concept of
            carbon sequestration, suggesting that it would be
            possible to capture CO_2 before it is emitted from
            power plants and dispose of it in the deep ocean.

            The United Nations Conference on Desertification
            convenes for the first time.

            The United States bans the use of CFCs as aerosol
            propellants in spray cans; the law goes into effect the
            following year.

1979   The first World Climate Conference convenes in
            Geneva. It results in the creation of the World Climate
            Data Program, the World Climate Applications Pro-
            gram, the World Climate Impact Studies Program,
            and the World Climate Research Program.

            The Convention on Long-Range Transboundary Air
            Pollution is adopted.

            The worst commercial nuclear accident in the United
            States occurs when equipment failures and human
            mistakes lead to a loss of coolant and partial core melt-
            down at the Three Mile Island reactor in Middletown,

1979
(cont.)
Pennsylvania. Thousands of people living near the plant leave the area before the 12-day crisis ends, during which some radioactive water and gases are released.

1980
Leo Donner and Ramanathan publish a paper that describes the greenhouse gas properties of methane and nitrous oxide and argues for their inclusion in climate models.

The Iran-Iraq War begins. Many Persian Gulf countries reduce oil production. Oil prices increase to unprecedented levels between 1979 and 1981, significantly raising the price of gasoline and other products. The average price of a gallon of gas in the United States reaches $1.41 ($3.22 in 2008 dollars).

The UNEP/ICSU/WMO Meeting of Experts on the Assessment of the Role of CO_2 on Climate Variations and Their Impact convenes in Villach, Austria, to investigate the effect of increasing greenhouse gas concentrations.

1982
WMO, UNEP, and ICSU representatives meet in Geneva and recommend that assessments of CO_2 be conducted every five years.

Global negotiations begin on a framework convention for the protection of the ozone layer.

1985
Analysis of ice cores taken from the Vostok station in East Antarctica shows that high CO_2 concentrations have coincided with high global temperatures throughout geologic history.

The Vienna Convention for Protection of the Ozone Layer is adopted.

Discovery of the Antarctic ozone hole is published in *Nature* by members of the British Antarctic Survey.

The World Climate Programme conference convenes in Villach, Austria, to discuss the role of carbon dioxide and other GHGs in climate variation. "As a result of the increasing concentrations of greenhouse gases," the conference proceedings assert, "it is now believed that in the first half of the next century a rise of global mean temperature could occur, which is greater than any in man's history."

Wallace Broecker speculates in an article in *Nature* that changes in precipitation or runoff could affect the formation of North Atlantic deep water and global ocean circulation. He posits that the change could be rapid and difficult for humans to adapt to or reverse.

1986   A numerical model developed by Mark Cane and Stephen Zebiak is the first to successfully predict an El Niño event.

Susan Solomon and colleagues publish an article detailing how chlorine atoms released from CFCs interact with ice crystals in polar stratospheric clouds and the polar wind vortex to create the ozone hole over Antarctica.

The worst accident in the history of nuclear power occurs at a plant in Chernobyl, Ukraine. Nearly 350,000 people are evacuated and eventually resettled, and 6.6 million are exposed to dangerous levels of radiation. The disaster intensifies opposition to nuclear power worldwide.

1987   Governments ask the WMO to work with UNEP to create an intergovernmental scientific group to provide a comprehensive assessment of climate science, the effects of increased concentrations of GHGs in the atmosphere, and related socioeconomic issues. This request leads to the creation of the Intergovernmental Panel on Climate Change (IPCC).

The Montreal Protocol on Substances That Deplete the Ozone Layer is adopted. The treaty mandates 50 percent cuts in the production and use of five CFCs

1987   and three other chemicals that destroy stratospheric
(cont.)   ozone.

U.S. government agencies are assigned the task of
developing a 10-year U.S. Global Change Research
Program.

1988   National Aeronautics and Space Administration
(NASA) scientist James Hansen testifies to Congress
that he has 99 percent confidence that a long-term
warming trend is underway.

The Conference on the Changing Atmosphere brings
together 300 policy makers and scientists from 46
countries to discuss the problems of atmospheric pollution. The conference statement recommends a
global goal of reducing greenhouse gas emissions by
20 percent of 1988 levels by 2005.

The UN General Assembly endorses actions of the
WMO and UNEP to create the IPCC. It commissions
the IPCC to produce a report on a variety of social
and scientific questions surrounding climate change.
The IPCC holds its first meeting in November.

1989   The U.S. Environmental Protection Agency (EPA)
publishes *The Potential Effects of Global Climate
Change on the United States,* which includes discussions of potential changes to natural systems and
biodiversity, sea levels, fisheries, agriculture, water
and energy demand, urban infrastructure, air quality, and health.

1990   The IPCC releases its First Interim Assessment Report. The most authoritative scientific assessment of
climate change at that time reports that global temperatures have increased by 0.3 to 0.6°C over the previous century.

The UN General Assembly takes official note of the
findings of the IPCC interim report, authorizes formal
negotiations on a global climate treaty, and calls for

the negotiations to conclude before the UN Conference on Environment and Development (scheduled for June 1992) takes place.

1991   The world's governments establish the Global Environment Facility to help developing countries fund projects and programs in the areas of biodiversity, climate change, international waters, and stratospheric ozone protection.

The United States hosts the first meeting of the global climate negotiations (officially known as the Intergovernmental Negotiating Committee for a Framework Convention on Climate Change).

1992   The UN Framework Convention on Climate Change (UNFCCC) is signed at the Earth Summit in June in Rio de Janeiro. The convention requires governments to take efforts to stabilize atmospheric concentrations of greenhouse gases at levels that would prevent dangerous human-induced interference with the climate system. Industrialized countries are required to adopt policies that aim to return emissions to 1990 levels. Though the United States signs the convention, the steps necessary to meet this goal are never taken.

Eliki Tajik and Takafumi Matsui present a model of the global carbon cycle that includes details about the flux and amount of carbon in all known reservoirs such as crust, mantle, atmosphere, ocean, and biosphere.

The U.S. federal government releases the *US National Action Plan for Global Climate Change*. The report includes federal and state, private, and cooperative measures, both legislative and administrative.

1993   Using a GCM, Syukuro Manabe and Ronald Stouffer show that increasing concentrations in atmospheric CO_2 could lead to drastic changes in, or a cessation of, current patterns of deep ocean circulation.

1993 *(cont.)*	Manabe and Stouffer, along with Thomas Delworth, suggest a link between interdecadal temperature variations such as the El Niño and Southern Oscillation, the North Atlantic Oscillation, and the thermohaline circulation.
1994	The UNFCCC enters into force.
1995	A 4,200-square-kilometer section of the Antarctic Larsen ice shelf disintegrates into hundreds of icebergs in January. Some scientists suggest that other ice shelves might be close to climatic limits and could disintegrate in the future.
	The First Conference of the Parties (COP 1) to the UNFCCC convenes in Berlin, launching new negotiations aimed at creating detailed and binding commitments for industrialized countries to reduce greenhouse gas emissions.
	The IPCC releases its Second Assessment Report. Written and reviewed by approximately 2,000 expert scientists, the report concludes that "the balance of evidence suggests that there is a discernible human influence on global climate" and notes the availability of "no-regrets" options to reduce greenhouse gas emissions.
1996	COP 2 for the UNFCCC convenes in Geneva. Governments attending the meeting, including the United States, adopt a declaration that accepts the findings of the IPCC report and calls for binding targets for greenhouse gas emissions.
1997	The U.S. Senate unanimously approves the Byrd-Hagel resolution, stating that the United States should not become party to any climate treaty that does not include timetables and emissions targets for developing countries or that causes harm to the U.S. economy. Though the resolution has no legal effect, it signals that the Senate will likely reject the Kyoto Protocol (then under negotiation).

Thomas Stocker and Andreas Schmittner publish research showing that the Atlantic thermohaline circulation is sensitive to both atmospheric CO_2 concentrations and the rate at which CO_2 is added. They suggest that the rapid addition of CO_2 could stop thermohaline circulation altogether.

The Kyoto Protocol to the Framework Convention on Climate Change is adopted in December. The treaty requires industrialized countries to reduce their collective GHG emissions by at least 5.2 percent below 1990 levels. Within this group, countries receive different national targets. The treaty also creates three important "flexible mechanisms" designed to lower the cost of reducing emissions while also accelerating the diffusion of relevant technology: the Clean Development Mechanism (CDM), Joint Implementation (JI), and International Emissions Trading. Details on how these mechanisms will operate are left to future negotiations.

1998   Michael Mann and colleagues publish an article containing what becomes known as the "hockey stick graph." The graph is a reconstruction of global average temperatures over the last millennium; it shows a record of little variation for hundreds of years marked by a dramatic and unprecedented increase in temperature beginning around 1900. Although initially disputed by several climate skeptics, it becomes an iconic symbol of man's influence on the climate system.

1999   Nineteen U.S. environmental organizations petition the U.S. EPA to promulgate regulations on greenhouse gas emissions from automobile tailpipes, claiming such action is required under the Clean Air Act.

2000   World leaders gathered at the UN Millennium Summit in New York City agree to pursue eight Millennium Development Goals aimed at eliminating conditions of extreme poverty around the world by 2015.

2000
(cont.)
The American Association for the Advancement of Science, an organization with more than 140,000 members, publishes a reported titled *An Atlas of Population & Environment*. The report concludes that unless measures are taken, atmospheric CO_2 concentrations will likely double from preindustrial levels, warming the world by 3°C and resulting in a high risk of extreme weather, rising sea levels, and the spread of pests and diseases to new regions.

2001
U.S. president George W. Bush announces in March that his administration will not support the Kyoto Protocol because it does not require emissions reductions from developing countries and would be too costly to implement. The announcement makes it clear that the United States will almost certainly not ratify the Kyoto Protocol before the end of its first commitment period in 2012.

The IPCC releases the Third Assessment Report, which concludes that evidence of human influence on the global climate is stronger than ever and presents a detailed picture of the regional effects of climate change. IPCC Working Group II (which focuses on impacts, adaptation, and vulnerability) notes that "observational evidence indicates that climate changes in the 20th century already have affected a diverse set of physical and biological systems."

Seventeen national and regional academies of science (Australia, Belgium, Brazil, Canada, the Caribbean, China, France, Germany, India, Indonesia, Ireland, Italy, Malaysia, New Zealand, Sweden, Turkey, and the United Kingdom) issue a joint statement supporting the IPCC report and affirming that the work of the IPCC represents the consensus of the international science community.

A statement offered on behalf of more than 270 Catholic bishops in the United States urges the public and government officials to recognize the seriousness of the global warming threat.

The Marrakech Accords are adopted at COP 7 in November, providing the necessary framework and details for the implementation of the emissions-reduction mechanisms created by the Kyoto Protocol.

Ramanathan and colleagues publish a study confirming how aerosols in air pollution cause cooling, which mask the impacts of greenhouse gases that would otherwise have raised temperatures even more. Recent air pollution control measures have limited the amount of aerosols in the atmosphere, ironically reducing the role they had been playing in counteracting some of the warming impacts of greenhouse gases.

2002   President Bush releases the *US Global Climate Change Policy Book,* announcing a voluntary commitment to reduce greenhouse gas intensity (the ratio of greenhouse gas emissions to economic output) by 18 percent in 10 years as an alternative to mandatory absolute reductions under the Kyoto Protocol. Critics argue that because the measures are both voluntary (unenforceable) and intensity based, they are unlikely to lead to substantial emission reductions.

The 70th Annual U.S. Conference of Mayors calls for state governments and the federal government to provide new resources to implement greenhouse gas emission reduction programs.

The Sierra Club, Greenpeace, and others bring a lawsuit against the U.S. EPA for failing to respond to their 1999 petition for regulation of CO_2 emissions from vehicles under the requirements of the Clean Air Act.

2003   Hernán Angelis and Pedro Skvarca find evidence that the collapse of ice sheets could trigger more rapid glacial outflow and enhance sea level rise. Their work is based on observations of Antarctic glaciers surrounding the site of the former Larsen-B ice shelf.

A record summer heat wave strikes Europe, causing estimated crop losses of approximately US$12.3 billion and nearly 35,000 deaths.

2003
(cont.)
A statement issued by the American Meteorological Society concludes that clear evidence exists that the mean annual temperature at the Earth's surface has been increasing for the past 200 years.

The EPA responds to the lawsuit brought by the Sierra Club and others by publishing a Notice of Denial of the petition to pursue rulemaking for CO_2 under the Clean Air Act. Led by Massachusetts, the attorneys general of 13 states file a petition with the U.S. District Court of Appeals requesting review of the EPA's Notice of Denial. The Sierra Club and Greenpeace also ask for a review, as the original plaintiffs. The court consolidates the cases into *Massachusetts et al. v. Environmental Protection Agency.*

The Chicago Climate Exchange (CCX) opens, providing continuous electronic trading of greenhouse gas emission allowances among members of the exchange. Members trade credits generated under a voluntary but legally binding cap-and-trade program administered by the CCX.

The American Geophysical Union, which represents 45,000 scientists from 140 countries, releases a statement concluding that human activities are increasingly altering the Earth's climate and that the unprecedented increases in greenhouse gas concentrations, together with other human influences on climate, constitute a real basis for concern.

2004
The U.S. Climate Change Science Program issues a report to the U.S. Congress stating that the global temperature increases observed in the latter half of the 20th century can be replicated by models only if human influences are included in the models.

NASA Goddard Institute for Space Studies Director Hansen reveals efforts by the current Bush administration to censor information that he and other government scientists publish or say publicly regarding climate change.

The Energy Action Coalition forms to coordinate the activities of North American youth groups engaged in climate change activism. The group organizes Energy Independence Day in October which includes more than 200 local events and collects 27,000 signatures from young people calling for drastic changes to the energy system.

Russia ratifies the Kyoto Protocol. Now 126 countries, representing 55 percent of total 1990 CO_2 emissions from developed countries, are party to the protocol. Under the terms of the treaty, reaching the 55 percent threshold allows the protocol to enter into force 90 days after Russia's ratification, despite the United States' nonparticipation.

Eight Arctic countries (Canada, Denmark, Finland, Iceland, Norway, Russia, Sweden, and the United States) and six indigenous peoples' organizations release the results of an unprecedented four-year scientific study of the Arctic, *The Arctic Climate Impact Assessment*. The report concludes that "the Arctic is now experiencing some of the most rapid and severe climate change on Earth."

The journal *Science* publishes an analysis of 928 abstracts randomly selected from a sample of 11,000 scientific research articles containing the keywords "global climate change" published in peer-reviewed scientific journals between 1993 and 2003. The article finds that none disagreed with the general consensus that the Earth's climate is changing and that most of the observed warming of the last 50 years is due to human activities.

Eight U.S. states and New York City bring a lawsuit against five major electric utilities citing a public nuisance caused by the carbon emissions from power plants as a contributor to global warming. The suit asks for emissions abatement.

2005   The Kyoto Protocol enters into force.

2005
(cont.)

The European Union Emissions Trading System begins operating as the largest multicountry, multisector greenhouse gas emission trading scheme worldwide.

The National Academies of Science of the Group of Eight (G8) countries plus China, Brazil, and India issue a statement declaring the consensus on climate change science is more than sufficient to justify prompt action and calling on leaders to acknowledge the threat of climate change.

The U.S. Senate passes a nonbinding Sense of the Senate resolution in June stating that "GHGs accumulating in the atmosphere are causing average temperatures to rise at a rate outside the range of natural variability" and that "there is a growing scientific consensus that human activity is a substantial cause."

President Bush signs the first national energy legislation in the United States in more than a decade. The bill does not acknowledge climate change but focuses on promoting energy self-sufficiency.

Hurricane Katrina hits the Gulf Coast in August, devastating communities in Louisiana and Mississippi and killing several thousand people. In addition, the hurricane causes extensive damage to the oil infrastructure in the Gulf of Mexico, leading to gas shortages and price increases in the United States. It is the most expensive and destructive natural disaster to occur in the United States. Media coverage includes extensive discussion of possible links between climate change and hurricane intensity, raising public awareness of climate change.

China announces a plan to increase the use of renewable energy from 7 to 15 percent of total energy production over the next 15 years. The plan includes $180 billion in investments and involves the replacement of 10 million tons of petroleum with renewable energy annually.

The Energy Action Coalition launches the Campus Climate Challenge through which groups working on college and high school campuses across North America work to reduce their schools' greenhouse gas emissions to zero.

Thirteen U.S. national and 122 local Jewish public affairs agencies issue a consensus statement on environmental issues that includes support for reducing greenhouse gas emissions in the United States and other industrialized countries.

The first Meeting of the Parties to the Kyoto Protocol (MOP 1) convenes in conjunction with the COP 11 of the UNFCCC.

The people of Papua New Guinea's Carteret Islands become the first environmental refugees evacuated because of climate change.

States and territories in Australia develop a nationwide emissions-trading scheme despite lack of federal government action.

Seven Northeast U.S. states create the Regional Greenhouse Gas Initiative, which uses a cap-and-trade system to stabilize CO_2 emissions from the region's power plants.

The U.S. Department of Energy launches the FutureGen Alliance with the goal of producing a zero-emission coal-fired power plant within 15 years.

2006   NASA reports that 2005 is the warmest year since record keeping began more than a century ago. The years 1998, 2002, 2003, and 2004 stand as the second-, third-, fourth-, and fifth-warmest years on record, respectively.

The Asia-Pacific Partnership on Clean Development and Climate is launched. As a result of the partnership, Australia, India, Japan, China, South Korea, and

2006
(cont.)

the United States agree to cooperate on developing and transferring technology to reduce greenhouse gas emissions.

More than 90 influential evangelical Christians in the United States endorse an environmental action plan which argues that the need to address climate change is urgent and calls on the Bush administration to impose mandatory limits on greenhouse gas emissions.

The U.S. House Interior Appropriations Subcommittee and the Senate Foreign Relations Committee adopt nonbinding resolutions acknowledging climate change as a serious problem.

Temperature Trends in the Lower Atmosphere, a report by the U.S. Climate Science Program, concludes that patterns of warming in the low and middle atmosphere are consistent with predictions of climate change and cannot be explained by natural processes alone.

The California state legislature passes the California Global Warming Solutions Act of 2006, requiring a 25 percent reduction in statewide CO_2 emissions by 2020. California is the twelfth-largest emitter of CO_2 in the world.

An Inconvenient Truth, a documentary film and part of former U.S. vice president Al Gore's public education campaign on climate change, premieres at the Sundance Film Festival.

The Institutional Investor Group on Climate Change, a forum for collaboration between very large institutional investors (such as state pension funds), issues a consensus statement accepting the broad scientific consensus on climate change and pledging to support efforts to develop solutions that avoid the risks of dangerous climate change.

German chancellor Angela Merkel announces that she will make climate change a central focus of Germany's 2007 European Union (EU) presidency.

The United Kingdom releases the *Stern Review of the Economic of Climate Change.* The report concludes that the costs of failing to prevent climate change will be very high; it also argues that the cost of reducing greenhouse gas emissions would be relatively small over the long term, as little as 1 percent of global gross domestic product (GDP) per year by 2050.

COP 12/MOP 2 convenes in Nairobi. Negotiators agree that the United Nations should establish an adaptation fund to assist developing countries in adjusting to climate change.

2007   The IPCC releases its massive Fourth Assessment Report. The report, which is released in stages throughout the year, confirms that climate change is occurring, that the human contribution to this change is unequivocal, and that impacts are already apparent and will increase as temperatures rise. In addition, the report concludes that mitigation is still possible if policies are rapidly put in place to reduce emissions and that many of these policies could be quite cost efficient.

The Global Roundtable on Climate Change releases a consensus statement supported by more than 100 major corporations and organizations calling for strong international policy to limit greenhouse gas emissions.

The U.S. Climate Action Partnership, a group of U.S.-based multinational corporations and environmental advocacy groups, calls on the U.S. government to enact strong limits on greenhouse gas emissions.

Combat Climate Change, a coalition of largely Europe-based businesses, issues a statement committing the signatories to nine principles regarding the need to combat climate change.

The European Union Parliament adopts a resolution calling for greenhouse gas emission reductions to limit the rise in global temperature to 2°C by 2100.

2007
(cont.)
The U.S. Supreme Court rules in favor of the plaintiffs in *Massachusetts et al. v. EPA,* requiring the U.S. EPA to revisit the question of whether CO_2 should be regulated under the Clean Air Act.

An Inconvenient Truth wins the Academy Award for best documentary.

In April, the inflation-adjusted average price of a gallon of gasoline reaches $3.32, exceeding a previous (inflation adjusted) all-time high reached in 1981.

In a worldwide call to action to mitigate climate change, the Live Earth concert stages 24 hours of live music on seven continents, featuring more than 150 of the world's most popular musicians.

President Bush proposes a "new global framework" for addressing climate change that includes a call for the world's top 15 emitters to form an agreement on the mitigation of greenhouse gases. Though the proposal marks the first time Bush has publicly accepted the ideal of global targets, the proposal is criticized for avoiding mandatory targets and for potentially undermining the existing UN and UNFCCC negotiations aimed at creating a successor agreement to the Kyoto Protocol.

The UNEP Finance Initiative, a large group of financial-sector companies that works together on the sector's role in responding to climate change, issues a statement calling on G8 governments to support immediate, deep cuts in carbon emissions.

The G8 summit convenes in Heiligendamm, Germany. President Bush reportedly blocks German chancellor Merkel's proposal for the G8 to endorse 50 percent reductions in greenhouse gas emissions by 2050.

The IPCC and Gore are awarded the 2007 Nobel Peace Prize for their efforts to create and disseminate knowledge about climate change.

Power Shift 2007, the first national youth conference on climate change, brings together 5,000 students at the University of Maryland in November to learn about climate change and to lobby members of Congress for legislation to address it.

In December, Congress passes new energy legislation that includes significant provisions to increase the energy efficiency of cars; light trucks; buildings; and home appliances, including dishwashers, dehumidifiers, lamps, residential boilers, walk-in freezers, and lightbulbs.

Australia ratifies the Kyoto Protocol, leaving the United States as the only major industrialized country outside the treaty.

COP 12/MOP 3 convenes in Bali, Indonesia. After two weeks of intensive talks, delegates agree on the Bali Roadmap, which sets a framework for concluding negotiations on the next phase of a global climate treaty in 2009. Parties also agree to establish an Adaptation Fund to assist developing countries in coping with the realities of a changing climate; it is funded by a tax on the sale of certified emission reductions in the CDM.

2008   Focus the Nation, a national "teach-in" on climate change, is held in January in the United States. More than 1,500 college campuses as well as a variety of civic organizations hold an unprecedented number of events and seminars exploring climate change issues.

In March, leaders for the 27 EU nations announce their intention to enact legislation mandating a 20 percent cut (from 1990 levels) in greenhouse gases by 2020 and requiring that 20 percent of total energy consumption come from renewable energies. The EU states that it would consider reducing GHG emissions by 30 percent if other developed countries also took significant action.

2008
(cont.)
Japan hosts the Fourth Ministerial Meeting of the G8 Gleneagles Dialogue on Climate Change, Clean Energy and Sustainable Development. The world's top greenhouse-gas-emitting countries discuss what steps should be taken after obligations to the Kyoto Protocol expire at the end of 2012. Japan pushes a "sectoral" approach—including setting energy efficiency goals for each industry—but this strategy is met with skepticism from developing countries.

Yale University releases a study and companion Web site on the economic impacts of reducing carbon emissions. The study examines thousands of policy simulations from 25 economic models used to predict the economic impacts of reducing U.S. carbon emissions. It shows that under the most pessimistic assumptions, a 40 percent reduction of GHGs by 2030 would still result in national GDP growth of 2.4 percent a year.

The World Glacier Monitoring Service releases data showing record glacial mass loss. Data indicate that the average rate of melting on 30 glaciers in nine mountain ranges more than doubled between 2004 and 2006. The biggest losses were recorded in the Alps and the Pyrenees mountain ranges of Europe.

A 160-square-mile chunk of sea ice in western Antarctica, about seven times the size of Manhattan, suddenly collapses in March, putting an even greater portion of ice at risk.

The Alliance for Climate Protection begins a three-year, $300 million advertising campaign to recruit 10 million individuals to advocate for policy and legislation to cut greenhouse gas emissions.

Scientists at the Mauna Loa observatory in Hawaii report that CO_2 levels in the atmosphere have reached a record high of 387 ppm, up almost 40 percent since the Industrial Revolution and the highest for at least the last 650,000 years.

Gasoline prices exceed $4 (in 2007 dollars) a gallon in the United States.

An international research team publishes a comprehensive study documenting the ways in which man-made climate change has already altered the behavior of thousands of plants, animals, rivers, glaciers, and other natural systems.

The U.S. Department of Agriculture releases "The Effects of Climate Change on Agriculture, Biodiversity, Land, and Water Resources." The report concludes that climate change is already impacting the nation's ecosystems and agricultural lands in significant ways, and those alterations will likely accelerate in the future, in some cases dramatically.

The Lieberman-Warner Climate Security Act, a proposal to establish a greenhouse gas emissions-trading scheme that would have covered an estimated 87 percent of emissions, as well as other emissions reduction measures, fails to win a key procedural vote and is withdrawn from consideration in 2008. Supporters express hope that the four full days of debate and a very close Senate vote indicate a strong climate bill will emerge from Congress in 2009.

A new consensus assessment by the National Intelligence Security Council—with input by all 16 U.S. intelligence agencies—finds that global climate change can trigger humanitarian disasters and political instability, which will seriously affect U.S. national security.

Japan hosts the G8 summit in July where the leaders of 16 major economies, including the United States, agree to cut emissions by 50 percent by 2050.

Australia introduces a national greenhouse gas emissions-trading system as part of the new Carbon Pollution Reduction Scheme, which seeks to begin reducing Australia's greenhouse gas emissions toward its Kyoto target.

2008
(cont.)   Ten northeastern U.S. states hold the first auction of carbon emissions allowances under the Regional Greenhouse Gas Initiative to curb greenhouse gases from the region's power plants.

5

Biographical Sketches

Svante Arrhenius (1859–1927)

Svante Arrhenius was the first scientist to investigate the relationship between CO_2 in the atmosphere and the temperature at the surface of the Earth; he was also the first to ask whether human activity could affect this relationship. Born in Vik, Sweden, Arrhenius exhibited an early affinity for mathematics. As a student at the Stockholm Academy of Sciences, he studied the conductivity of dissolved substances and developed the theory that ions—positively or negatively charged particles that split off from the molecules in solution—were responsible for conducting the electrical charge through the solution. When he presented his dissertation on this topic in 1884, the faculty was unimpressed, awarding his paper the lowest possible passing grade.

However, scientists at other institutions more readily understood the significance of Arrhenius's work. He was appointed professor of physics at Stockholm University in 1891. In 1903, he was awarded the Nobel Prize in chemistry for his work on the ionization of electrolytes.

Arrhenius also investigated whether variations in carbon dioxide had contributed to prehistoric ice ages. He undertook a series of laborious calculations to quantify the potential impact of carbon dioxide on temperatures at the surface of the Earth. In his seminal article "On the Influence of Carbonic Acid [CO_2] in the Air upon Temperature on the Ground," published in 1896, Arrhenius demonstrated that a doubling of atmospheric CO_2 could lead to a temperature increase of between 4 and 6°C, with the greatest variation at the poles. He also showed that a reduction

155

by half of atmospheric CO_2 levels could reduce temperature to a similar extent. While Arrhenius did not foresee that human activity would create sufficient CO_2 emissions to threaten the stability of the climate system, he did pose the question and thus set the stage for subsequent research on the subject.

Bert R. Bolin (1925–2007)

A leading atmospheric scientist, Bert R. Bolin played a central role in international efforts to increase global knowledge, awareness, and consensus regarding climate change. From 1988 until 1998, Bolin served as the first chair of the Intergovernmental Panel on Climate Change (IPCC), the body established by the world's governments to provide comprehensive, authoritative assessments of current knowledge of the natural science, impacts, and policy dimensions of climate change. He authored more than 170 articles, including many early publications pointing to the disturbance of the carbon cycle by anthropogenic CO_2.

Born in Stockholm, Sweden, Bolin began his career studying the mathematics of atmospheric circulation and developing numerical equations for weather prediction. In 1959, he and his colleague Erik Eriksson published the first clear explanation of the limits of ocean uptake of CO_2 (Bolin and Eriksson 1959). Their research showed that while the top layer of the ocean absorbs significant quantities of CO_2, most of it evaporates before being mixed into the deeper ocean and ends up in the atmosphere. Based on this research, they projected that atmospheric CO_2 could increase by 25 percent by the end of the 20th century.

In 1968, Bolin was elected chair of the organizing committee for the Global Atmospheric Research Program, a joint project of the World Meteorological Organization and the International Council of Scientific Unions to coordinate international research on the atmosphere by gathering data on a global scale.

Bolin served on the Advisory Group on Greenhouse Gases, the precursor group to the IPCC, and later chaired the IPCC from its launch in 1988 until 1998. Among the important challenges he undertook in that role, Bolin served as IPCC's liaison to the negotiators of the United Nations Framework Convention on Climate Change (UNFCCC), attempting to ensure that the convention reflected the most authoritative science. His leadership helped to advance the international cooperation critical to understanding and responding to global climate change, and he has been recognized with many awards, including the American Meteorological Society's Carl-Gustaf Rossby Prize (1984) and the Blue Planet Prize (1995).

Wallace S. Broecker (b. 1931)

Wally S. Broecker is Newbury Professor of Earth and Environmental Sciences at Columbia University and one of the world's preeminent geophysicists and paleoclimatologists. His work on ocean chemistry revealed many biological, chemical, and physical processes that govern the behavior of carbon dioxide in the oceans and the interactions of oceanic carbon dioxide with the atmosphere. His research has helped scientists to understand both ancient climates and how atmospheric changes will affect the climate of the future.

Born in Chicago, Broecker attended Columbia University, earning an undergraduate degree in physics (1953) and an MA (1954) and a PhD in geology (1958). He has spent nearly his entire career at Columbia, in the Department of Earth and Environmental Sciences (formerly the Department of Geology) and the renowned Lamont-Doherty Earth Observatory (LDEO).

In his early research, Broecker used the weathering of different forms of carbon atoms to explore past climates. In 1966, he published data from studies of ocean sediments showing that in the distant past, climate had changed much more abruptly than previously believed (Broecker 1966). In 1970, with his colleague Jan van Donk, Broecker published a groundbreaking paper that used analysis of sediment samples as evidence of a 100,000-year cycle of glaciation, confirming a theory developed in the early 20th century by Milutin Milankovich, who proposed that cycles of glacial advance and retreat were caused by fluctuations in solar radiation resulting from variations in Earth's orbit around the sun (Broecker and van Donk 1970).

In 1985, Broecker and his colleagues outlined what they called the "great ocean conveyer belt"—the mechanism by which the ocean carries heat from the tropics to the poles (Broecker, Peteet, and Rind 1985). Taken together with new evidence of rapid climate change in ice cores, Broecker's results presented a new possibility for abrupt climate change, as a sudden change in the conveyer-belt circulation could cause drastic changes to global climate. This discovery was an important advance in understanding the risks associated with accumulation of carbon dioxide in the atmosphere.

Broecker has received many awards and honors, including membership in the National Academy of Sciences, the American Academy of Arts and Sciences, and both the American Geophysical Union and the European Geophysical Union. He has also received the Maurice W. Ewing Medal (1979) and the Roger Revelle Medal (1995) from the American Geophysical Union; the Arthur L. Day Medal (1984) and Don J. Easterbrook Distinguished Scientist Award (2000) from the Geological Society of America; the A. G. Huntsman Award for Excellence in the Marine Sciences (1985); the Urey Medal from the European Geophysical Union (1986); the Alexander Agassiz Medal from the National Academy of Sciences (1986); the V. M. Goldschmidt Award from the Geochemical Society (1986); the Blue Planet Prize for achievement in global environmental research (1996); and the United States' highest scientific honor, the National Medal of Science (1996).

Guy Stewart Callendar (1898–1964)

Guy Stewart Callendar is best known for his pioneering work constructing records of past temperatures and CO_2 levels and for proposing a causal link between the two, opening a critical line of thinking. Born in Canada in 1898, Callendar spent his childhood in and around London. He studied engineering at the Royal College of Science and then worked with his father, Hugh Longbourne Callendar, researching the properties of steam. He also became an expert in the properties of infrared radiation.

Callendar's strong personal interest in meteorology led him to begin gathering historical records of temperature and CO_2 from around the world. When he discovered that both were trending upward, he formulated the hypothesis that rising CO_2 could be inducing a rise in temperatures. In a 1938 article entitled "The Artificial Production of CO_2 and Its Influence on Temperature," Callendar argued that CO_2 from fuel combustion and other human activities was accumulating in the atmosphere and causing temperatures to rise (Callendar 1938). Using a simple model for increased re-radiation in the lower atmosphere caused by CO_2 accumulation, he theorized that a doubling of CO_2 would raise global temperatures by 2°C.

This article was the first in a series elaborating on the relationship between CO_2 and temperature. Callendar's work revived the CO_2 theory of global warming, which had been suggested in the 19th century through the work of Svante Arrhenius and John Tyndall but had largely disappeared by Callendar's time. Pioneering climate scientists Roger Revelle and Hans Suess later referred to the relationship between increased CO_2 emissions and temperature as the Callendar effect, demonstrating the historical importance of Callendar's research to the study of climate change. At the time of his death in 1964, Callendar was working on a prescient manuscript entitled "Carbon Dioxide and Climate Change."

Mark A. Cane (b. 1944)

Mark A. Cane and his colleague Stephen Zebiak developed the first computer models able to simulate and predict the El Niño–Southern Oscillation. El Niño is a periodic shift in tropical Pacific sea surface temperatures that has a major influence on weather and climate around the globe.

Born in Brooklyn, New York, Cane earned a BA (1965) and an MS (1966) at Harvard University and a PhD in meteorology at Massachusetts Institute of Technology (1975). He joined Columbia University and its Lamont-Doherty Earth Observatory research institute in 1984. Today, Cane is the G. Unger Vetlesen Professor of Earth and Climate Sciences in the departments of Earth and Environmental Sciences and Applied Physics and Mathematics; he also holds appointments at LDEO and the International Research Institute for Climate and Society (IRI).

In 1985, a model developed by Cane and Zebiak made the first physically based forecasts of El Niño. This was extremely important because, apart from the seasonal cycle itself, no other phenomenon influences short-term climate variability so profoundly. Over the years, the Cane-Zebiak model, as it is now known, has been the primary tool used by many researchers to enhance science's understanding of El Niño events (Chen et al. 1995).

Cane has also worked extensively to predict and mitigate the impact of El Niño on human activity. He was instrumental in the founding of IRI, whose mission is to use forecast abilities to mitigate the adverse socioeconomic and environmental effects of

El Niño and other climate variations. With this aim, IRI dissemi-
nates forecasts to assist farmers and government planners with
decisions about storing grain, planting crops, conserving water,
and building dams. It also develops and deploys strategies to
deal with climate risk.

Cane has written some 200 papers on a broad range of topics in oceanography and climatology and has served on numerous international and national committees. He is the recipient of
the Sverdrup Gold Medal of the American Meteorological Society (1992) and the Cody Award in Ocean Sciences from the
Scripps Institution of Oceanography (2003); he is also a fellow of
the American Meteorological Society, the American Association
for the Advancement of Science, the American Geophysical
Union, and the American Academy of Arts and Sciences. His
current research focuses on explaining variations in the paleocli-
mate record.

Eileen Claussen (b. 1945)

Eileen Claussen is president of the Pew Center on Global Climate
Change. She has been a prominent figure in public discussions of
climate change policy in the United States.

Claussen studied public policy at George Washington
University and the University of Virginia. In 1972, she joined
the newly created U.S. Environmental Protection Agency
(EPA), where she worked on hazardous waste disposal and
other issues. From 1987 to 1993, she was director of Atmospheric Programs at the EPA and led policy development on
many important issues, including ozone depletion, acid rain,
and end-user energy efficiency. She served as assistant secretary of state for the departments of Oceans and International
Environmental and Scientific Affairs during the 1990s, leading
policy development and implementation on international environmental issues, including stratospheric ozone depletion and
climate change.

Claussen left the government in 1998 to found the nonprofit Pew Center on Global Climate Change. Her long experience in the development and implementation of international
environmental policy is evident in the Pew Center's balanced
approach to the issue of climate change. The Pew Center con-

ducts research and produces policy proposals and analyses in consultation with corporations, governments, and scientists in order to develop solutions with broad-based support. Claussen and other Pew Center researchers frequently testify before the U.S. Congress and participate actively in the UN climate negotiations. Claussen and the Pew Center are a valuable voice for developing sensible climate policy both in the United States and internationally.

William Maurice "Doc" Ewing (1906–1974)

William Maurice Ewing has been described as the creator of the science of geophysics and was a great innovator in the study of geology, oceanography, and Earth science in general. In addition, Ewing's work, perhaps more than that of any other individual, laid the foundation for the revolutionary concept of plate tectonics.

Born in rural Texas, Ewing earned a BA in mathematics and physics (1926) and an MA (1927) and a PhD in physics (1931) from Rice University. During the 1930s, Ewing served as a professor of physics at the University of Pittsburgh and at Lehigh University, where he took the groundbreaking step of applying physics and chemistry to the study of geology. His first innovation was using sound waves to explore subsurface geology, which allowed geologists to explore deep layers beneath the ocean floor for the first time. Ocean sediments became an important source of information about past climate, a development that would not have been possible without Ewing's innovations.

Ewing was hired by Princeton University and the U.S. Geodetic Survey in 1934; in 1940 he moved his research group to the Woods Hole Oceanographic Institute in Massachusetts, where they adapted their methods for military purposes, developing tools to help submarines escape detection. After World War II, Ewing became a professor of geology at Columbia University.

By 1949, Ewing's many research breakthroughs helped convince Columbia University to establish the Lamont Geological Observatory (subsequently renamed the Lamont-Doherty Earth Observatory). Ewing served as the institute's first director, heading the organization for 25 years. Under his leadership, LDEO became one of the world's leading research institutions for the Earth sciences and the study of climate.

Jean-Baptiste Joseph Fourier (1768–1830)

Jean-Baptiste Joseph Fourier was the first person to conclude that the atmosphere played an important role in determining the temperature at the Earth's surface. Born in France in 1768 and educated by Benedictine monks, Fourier displayed strong skills in mathematics at an early age. He attended a teaching academy at the École Normale in Paris, and then taught mathematics at the Collège de France and the École Polytechnique.

In the 1790s, Fourier led a scientific expedition to Egypt under the command of Gen. Napoleon Bonaparte. He returned to France in 1801 and resumed teaching and research. In this period, he discovered that the movement of heat through solid objects could be described in a finite series of equations. His interest in heat conduction also led him to examine what factors determined the Earth's temperature.

In his seminal 1822 book, *The Analytical Theory of Heat*, Fourier suggested that the atmosphere played a critical role in warming the Earth's surface. He recognized that the Earth is heated by visible light from the sun, which easily penetrates the atmosphere. He also recognized that the Earth emits infrared radiation that it has absorbed from the sun and that some of the sun's heat is retained at the Earth's surface. If this were not the case, Fourier deduced, temperatures at the Earth's surface would be much lower than they actually were. Fourier attributed this warming to the gases of the atmosphere, comparing the effect to the warming of air in a glass-sided container and giving rise to the greenhouse analogy that persists to this day.

Albert Gore Jr. (b. 1948)

Albert Gore Jr., the 45th vice president of the United States, was one of the first American politicians to grasp the seriousness of climate change and call for a reduction in greenhouse gas emissions. His tireless work explaining and publicizing the issue early in the 21st century earned him the 2007 Nobel Peace Prize, one of the world's highest honors, which he shared with the IPCC.

Born in Washington, D.C., where his father, Albert Gore Sr., was a U.S. senator representing Tennessee, Al Jr. studied government at Harvard University, enlisted and served in the U.S. Army in Vietnam, worked as a journalist, and attended both law and divinity school before being elected to the House of Representatives in 1976. Gore represented his father's former congressional district in the House until 1984, when he won election to the Senate.

From the beginning of his political career, Gore advocated for increased government protection of the environment. In 1982, he cosponsored the first congressional hearings on global climate change, giving scientist James Hansen an opportunity to publicize early model results showing climate impacts from CO_2 accumulation (U.S. Congress 2007a). In 1992, Gore published the best-selling book *Earth in the Balance,* in which he describes the global ecological crisis and proposed solutions.

After unsuccessfully seeking the Democratic Party nomination for president in 1988 and 1992, Gore was elected vice president (sharing the ticket with Bill Clinton) in 1992. As vice president, Gore continued to work on environmental issues. Actively engaged in the negotiation of the Kyoto Protocol, Gore traveled to Kyoto in the final days of deliberations to try to break the deadlock. However, because the U.S. Senate had already declared that it would not allow the United States to become a party to any climate treaty that did not include targets for developing countries, the United States did not ratify the agreement.

Gore ran for president against George W. Bush in 2000, losing one of the closest, and perhaps the most controversial, elections in U.S. history. A series of voting irregularities in Florida led to vote recounts around the state that continued for weeks. The Bush campaign went to court to stop the recounts, and the U.S. Supreme Court eventually issued a decision halting the recount and effectively handing the election to Bush.

After the 2000 election, Gore dedicated himself to combating global climate change. He began a campaign to raise awareness of the issue, traveling the United States with a presentation on the science, politics, and economics of climate change. In 2006, Gore produced a book and a movie based on his presentation titled *An Inconvenient Truth,* both of which drew global attention and acclaim. The film became the third-highest grossing documentary ever released in the United States and won that year's Academy Award for best documentary.

James Hansen (b. 1941)

James Hansen is the director of the Goddard Institute of Space Studies (GISS), a unit of the National Aeronautics and Space Administration (NASA) that studies planetary systems, including the Earth system. Since the early 1980s, Hansen has been pivotal in advancing scientific understanding of climate change and increasing public awareness of the issue by actively promoting the scientific evidence of humanity's influence on climate and the probability that failure to address climate change will bring overwhelmingly negative consequences.

Born in Denison, Iowa, Hansen earned a BA in physics and mathematics (1963), an MA in astronomy (1965), and a PhD in physics (1967) from the University of Iowa. His dissertation examined the atmosphere of Venus. His focus shifted to the dynamics of Earth's atmosphere when he went to work at NASA, and he began running an early general circulation model to simulate the impacts of different factors, including the addition of greenhouse gases.

In 1981, Hansen was appointed director of GISS. That year, he and his colleagues published modeling results that closely replicated past climate, demonstrating the model's fundamental soundness. The model also showed that increased CO_2 would raise global average temperature by approximately 2.5°F by the year 2000 (Hansen et al. 1981). The article was one of the first to use the term *global warming* to describe the current phenomenon. In 1988, Hansen gave groundbreaking testimony to the U.S. Congress, stating that he was virtually certain that greenhouse gas emissions were causing climate change.

In 2005, after more than 30 years at NASA, Hansen began speaking out against the increasing politicization of the organization under the administration of George W. Bush. He reported that since 2004, officials at NASA headquarters had been reviewing his lectures, papers, and postings on the GISS Web site in an apparent attempt to censor the flow of scientific information regarding climate change (U.S. Congress 2007a, 2007b).

Hansen has published hundreds of scientific papers and has been featured on television and radio. He is a member of the National Academy of Sciences and has received numerous awards, including the John Heinz Environment Award and the Roger Revelle Award. Hansen continues to conduct important research and to speak out about the critical challenge of climate change and the importance of scientific freedom.

John Houghton (b. 1931)

Sir John Houghton, a pioneering climate scientist, has devoted his career to researching the impact of anthropogenic greenhouse gas emissions and has consistently urged global leaders to take action to address climate change. An evangelical Christian who advocates for the idea that science and religion are complementary, Houghton stresses that protection of the global environment is a religious and moral obligation.

Born in Dyserth, Clwyd, in Wales, Houghton began researching climate change as a student at Oxford University in the early 1960s; he became a professor of atmospheric physics there in 1976. Houghton developed remote sensing techniques that made it possible for satellites to gather information about the temperature and composition of the stratosphere. Scientists used these tools to monitor atmospheric CO_2, measurements that were critical to understanding global climate change.

In 1983, Houghton was named chief executive at the Met Office, the UK government meteorological agency. He was instrumental in founding the Met Office's Hadley Center, the center of climate change research in the United Kingdom and a center in climate modeling. Houghton was involved with the IPCC from its inception, serving from 1988 to 2002 as cochair of the physical science working group (Working Group I) and a lead editor of the assessment reports. The author of many books and articles, Houghton has spoken widely on climate change, even providing testimony to the U.S. Senate on the scientific basis for urgent action. He has also received many notable honors, including the 2006 Japan Prize and the 1999 International Meteorological Organization Prize.

Charles David Keeling (1928–2005)

Charles David Keeling pioneered the collection of accurate, constant measurements of atmospheric CO_2 concentrations. Keeling's measurements from the top of Mauna Loa volcano in Hawaii confirmed rising levels of CO_2 over the long term, provided evidence that this rise is driven by increasing emissions from human activities, and revealed a seasonal cycle of CO_2 reflecting the growth and decay of plant matter. The oscillating line graph, which shows steadily increasing concentrations of CO_2 in the atmosphere, has become one of the most recognizable images in modern science and is known as the Keeling Curve. Keeling's measurements fundamentally changed our understanding of the global system and formed a cornerstone of modern climate science.

Born in Scranton, Pennsylvania, Keeling earned a BA at the University of Illinois in 1948 and a PhD from Northwestern University in 1954, both in chemistry. In 1955, as a postdoctoral student in geochemistry at the California Institute of Technology, he embarked upon his first project to study factors regulating CO_2 levels in the atmosphere.

As part of his research, Keeling had hoped to accurately measure atmospheric CO_2 but could find no device suitable to the task. Accordingly, he designed his own equipment and techniques, which he refined and tested at locations throughout California until he began to record stable measurements in the most pristine of locations. He found that, when conducted away from pollution and vegetation, his measurements were remarkably consistent. This approach established the concept of "atmospheric background," which became important in studying the impacts of other atmospheric gases such as methane. Roger Revelle, the director of the Scripps Institute for Oceanography, secured funding from the International Geophysical Year to pay for a global survey of CO_2 levels and employed Keeling to establish monitoring stations atop Mount Mauna Loa and at a military base in the Antarctic. Keeling and Revelle believed that these locations were undisturbed enough to provide accurate measurements.

By December 1958, Keeling had collected his first full year of data, which indicated a slight increase in CO_2 over the course of the year. Despite funding difficulties, Keeling continued monitoring and, in 1962, published his first results, demonstrating that the baseline level of CO_2 in the atmosphere had increased. This research confirmed that the oceans were not taking up nearly as much of the CO_2 emitted from fossil fuel consumption as many scientists had previously assumed. His calculation of the fraction (the Keeling Fraction) of CO_2 that remains in the atmosphere after being emitted by fossil fuel combustion is still used in many climate models.

Keeling also worked on a number of aspects of carbon geochemistry, including carbon-cycle modeling. In 1996, he and colleagues showed that the amplitude of seasonal CO_2 variation in the Northern Hemisphere was increasing.

In recognition of his pioneering contributions to the advancement of climate science, Keeling received the National Medal of Science—the highest U.S. award for scientific research lifetime achievement—in 2002. Among many other notable honors, Keeling has also received the Maurice Ewing Medal of the American Geophysical Union (1991), the Blue Planet Prize (1993), the Tyler Prize (2005), and, in 1997, a special achievement award from then vice president Al Gore for his 40 years of research monitoring carbon dioxide at the Mauna Loa observatory.

Richard S. Lindzen (b. 1940)

A noted atmospheric scientist, Richard S. Lindzen earned his AB, SM, and PhD degrees from Harvard University. He has worked at the National Center for Atmospheric Research, the University of Chicago, and Harvard University. In 1983, he was appointed Alfred P. Sloan Professor of Meteorology at the Massachusetts Institute of Technology, a position he still holds today.

Lindzen has argued against the scientific consensus that climate change is caused mainly by human activities that introduce CO_2 into the atmosphere; he also opposes the idea that climate change poses a grave threat to the planet. Though he does not dispute that CO_2 levels and temperatures have risen over the past 100 years, he argues that these changes have not been conclusively linked to human activities nor shown to be harmful.

Lindzen maintains that the scientific community has greatly exaggerated the extent and dangers of climate change primarily to maintain government funding for individual research projects, which he believes would become scarce if the problem were not seen as urgent. Lindzen believes it is this financially motivated alarmism that has generated public concern and political movement toward regulation of greenhouse gas emissions and that restricting the use of fossil fuels is unwarranted.

Despite being a lead author on chapter 7 of the IPCC's Third Assessment Report, published in 2001, Lindzen is a vocal critic of the IPCC, calling it politicized and unreliable. Lindzen has also claimed that skeptical views such as his do not receive serious attention but attract threats and intimidation and make it difficult to obtain research funding.

The author of more than 200 scholarly papers, Lindzen has received the Macelwane Medal of the American Geophysical

Union and the Meisinger and Charney awards of the American Meteorological Society. While his distinguished credentials differentiate Lindzen from many outspoken skeptics, it remains to be seen if he will be best remembered for his outspoken doubts on climate change, increasingly overwhelmed by global opinion and physical evidence, or his previous achievements.

Bjørn Lomborg (b. 1965)

A statistician, Bjørn Lomborg is the author of the best-selling book *The Skeptical Environmentalist.* The book argues that many of the most well-publicized claims of environmentalists—including the rate of species extinction, the scale and cost of climate change impacts, and the extent of the environmental threats to human health—are exaggerated. The book's primary argument is that, in comparison to other global issues—particularly poverty—environmental problems are not as bad as environmental activists make them out to be. In Lomborg's view, environmental groups are motivated by self-interest to inflate their claims of degradation.

After the book's English-language publication in 2001, many members of the international scientific community accused Lomborg of distortion, misrepresentation, and highly selective use of data in his book. Though the Danish Ministry Agency for Science, Technology and Innovation Committee on Scientific Dishonesty investigated his work and cleared him of charges of intentional scientific dishonesty, many scientists remain harshly critical of Lomborg's work. In the view of these scientists, *The Skeptical Environmentalist* distorts the current state of the literature, uses evidence selectively, and deliberately courts controversy by making outlandish claims without sufficient support.

In 2004, Lomborg organized a group of economists (dubbed the Copenhagen Consensus) and tasked them with prioritizing responses to the world's biggest challenges. Shortly thereafter, Lomborg assembled a group of UN ambassadors to tackle the same topic. The two groups identified a set of priorities that Lomborg detailed in two recent books: *Global Crises, Global Solutions* and *How to Spend $50 Billion to Make the World a Better Place.* Taking a similar position to that of *The Skeptical Environmentalist,* Lomborg argues in these later volumes that investing funds in climate change mitigation is not an efficient strategy for global prosperity.

Critics have responded to these claims by arguing that investing in climate change mitigation does not have to be done to the exclusion of other investments. They have also argued that Lomborg has underestimated the economic, social, health, and environmental impacts of climate change, overestimated the costs of reducing emissions, and overlooked the ancillary benefits associated with reducing reliance on fossil fuels, developing alternative energy, and adapting to climate variability.

Lomborg holds an MA from the University of Aarhus and a PhD in political science from the University of Copenhagen.

Michael Mann (b. 1965)

Born in Amherst, Massachusetts, Michael Mann studied physics and applied math at the University of California, Berkeley (BS 1989), and later earned a PhD in geology from Yale University (1996). Mann has taught at the University of Massachusetts and the University of Virginia; in 2005, he became the director of the Earth System Science Center at Pennsylvania State University.

In recent years, Mann has become widely known for a chart referred to as the "hockey stick" graph. To produce the graph, which shows a sharp increase in the Earth's temperature during the last 50 years after centuries of gradual change, Mann and his colleagues relied on proxy data, including tree rings, ice cores, and coral, to provide evidence of past temperatures. He also employed the instrumental temperature record.

First published by Mann and colleagues in 1998, Mann's results showed that temperatures rose dramatically in the 20th century after centuries of relative stability (Mann, Bradley and Hughes 1998, 1999). This hockey-stick–shaped graph became iconic when it was prominently featured in the 2001 IPCC Third Assessment Report.

Although many studies had come to similar conclusions, Mann's graph became highly politicized, in part because of its inclusion in the IPCC report. To many, it was seen and used as clear evidence of the urgent need for action on climate change. Others, however, questioned its veracity. For instance, two economists published a response that challenged the statistical methods used to calibrate the proxy data. Later, in an act widely seen as political intimidation, Rep. Joe Barton (R-TX) demanded information about the study and about Mann's government research grants.

While a study by the National Academy of Sciences affirmed Mann's methods and conclusions, Mann responded to what he and others saw as a disinformation campaign waged by those opposed to climate change mitigation in the United States by founding the Web site Realclimate.org. This Web site helps inform the public about the complexities of climate science and policy.

Angela Merkel (b. 1954)

Angela Merkel is the first woman to serve as chancellor of Germany. She is also the first person born in what was once East Germany to assume the chancellery of the reunited state. Born in Hamburg, Merkel grew up in Templin, 50 miles north of Berlin in the former German Democratic Republic. She studied physics at the University of Leipzig, received a PhD based on her dissertation on quantum chemistry, and started her career in research.

After the fall of the Berlin Wall in 1989, Merkel became interested in the growing democracy movement and entered politics. She was elected to the Bundestag (parliament) in Germany's first post-reunification elections. She later became minister for Women and Youth and, in 1994, minister for Environment and Reactor Safety. Since 2000, she has served as the head of the Christian Democratic Union. In 2005, she took control of a coalition between this and two other parties. Merkel assumed office as the chancellor of Germany in November 2005.

Merkel has made climate change a priority of her administration, calling it the "greatest challenge of the 21st century" on more than one occasion. She was a driving force behind the European Union's (EU) *20 20 by 2020* climate initiative, which established a Union-wide goal of cutting carbon emissions 20 percent and using 20 percent renewable energy by 2020. Merkel has pushed for more ambitious greenhouse gas reductions by 2050.

In 2007, Merkel used Germany's dual role as president of the EU and head of the Group of Eight (G8) to promote stronger climate policy in Europe, among the major industrialized countries, and across the world. Her work has included urging the United States to take action, crafting strong European positions in the global negotiations, and promoting climate issues in bilateral EU summits with China, India, and Latin America.

Mario Molina (b. 1943)

Mario Molina played a central role in the critical discovery that man-made chlorofluorocarbons (CFCs) threatened the Earth's ozone layer, which helps shield the planet from ultraviolet radiation. For this work, Molina and his colleague F. Sherwood Rowland won the Nobel Prize.

Born in Mexico, Molina earned his PhD in physical chemistry from the University of California (UC), Berkeley. As a postdoctoral fellow under F. Sherwood Rowland at UC Irvine, Molina began his investigation of CFCs, industrial chemicals that were used extensively at that time as refrigerants, as aerosol propellants in spray cans, and as solvents. In 1974, Molina and Rowland published a groundbreaking paper in *Nature* explaining that CFCs, which were then thought to be inert, could break down, releasing chlorine atoms into the stratosphere and depleting the stratospheric ozone (Rowland and Molina 1974).

This paper led to years of research and debate and, eventually, to a global consensus on the need to stop production and use of CFCs and other ozone-depleting substances. The resulting treaty, the 1987 Montreal Protocol and the series of amendments that have strengthened it, is widely regarded as having averted a global environmental crisis. Molina's and Rowland's research began a process that not only led to the protocol but also alerted scientists and the public to the fact that human activity could have vast, unforeseen impacts on the global atmosphere, knowledge that was essential for understanding and addressing climate change. For their critical contribution in exposing and publicizing the dangers of CFCs, Molina and Rowland (along with Paul Crutzen, who had done earlier work on the ability of chlorine atoms to destroy ozone) received the Nobel Prize in chemistry in 1995.

Molina has worked at UC Irvine, the Jet Propulsion Laboratory, and the Massachusetts Institute of Technology; he currently holds a position at the University of California at San Diego, where his research focuses on the chemical properties of atmospheric pollutants. Molina also heads a center for strategic study of energy and the environment in Mexico City, where he works on issues relating to air quality and global change, focusing on the effect of particles on clouds and climate.

William Nordhaus (b. 1941)

William Nordhaus is Sterling Professor of Economics at Yale University. He is best known for his pioneering work on modeling the economic impacts of biophysical changes, including climate change. His work has been central to efforts to understand the profound long-term economic costs of climate change.

Born in Albuquerque, New Mexico, Nordhaus received his undergraduate degree from Yale University in 1963 and a PhD in economics from the Massachusetts Institute of Technology in 1967. He joined the faculty of Yale University and was tenured as a full professor of economics there in 1973. From 1977 to 1979, Nordhaus served as a member of the President's Council of Economic Advisers under President Jimmy Carter. He then returned to Yale and resumed teaching, also serving as provost of the university from 1986 to 1988.

The author of numerous books, articles, and other publications, Nordhaus joined with Paul Samuelson to develop the classic textbook *Economics,* which had its 18th edition published in 2005. The recipient of many honors, he has been elected a member of the National Academy of Sciences, the Swedish Academy of Engineering Sciences, a fellow of the American Academy of Arts and Sciences, and a fellow of the Econometric Society.

Nordhaus's research focuses on the relationship between economic growth and natural resources, including constructing integrated economic and scientific models to determine the most efficient path for coping with climate change. In 1974, while spending a year at the International Institute of Applied Systems Analysis in Vienna, he and a group of colleagues developed the first economic model for climate change. Nordhaus continued this work, subsequently developing the influential Regional Integrated Model of Climate and the Economy (RICE) and the Dynamic Integrated Model of Climate and the Economy (DICE). The most current modeling approach to climate change is the DICE-2007 model, the fifth major version. While many of the equations and details have changed over time, the basic modeling philosophy remains unchanged: to incorporate the latest economic and scientific knowledge and to capture the major elements of the economics of climate change in as simple and transparent a fashion as is possible.

In the late 1990s, Nordhaus began to work on models that could incorporate data on a smaller scale, enabling more accurate assessments of the impacts of various changes. His new models used geophysical data to reflect the relationship between small-scale climatic changes and economic activity. To further this research, Nordhaus started the G-Econ project, which was responsible for collecting the necessary data to allow the models to work on the new scale. With the geophysical scale in place, model simulations of the impacts on economic activity caused by climate change revealed more negative effects than previously thought.

Rajendra K. Pachauri (b. 1940)

Rajendra K. Pachauri became chairman of the IPCC in 2002 and oversaw the drafting and publication of the panel's Fourth Assessment Report in 2007. In this capacity, he shared the 2007 Nobel Peace Prize, on behalf of the IPCC, with Albert Gore Jr.

Born in Nainital, India, Pachauri earned two doctorate degrees—one in industrial engineering and the other in economics—from North Carolina State University. Pachauri served on the faculty at North Carolina State before returning to India. He joined the Administrative Staff College of India, Hyderabad, and went on to become the director of the Consulting and Applied Research Division. In 1981, he accepted the position of director at the Tata Research Institute, now known as The Energy and Resources Institute (TERI). He accepted the new position of director general of TERI in 2001 and continues to work in that capacity.

Pachauri became involved with the IPCC early in its history, serving as lead author of chapters produced by Working Groups II and III for the panel's 1995 Second Assessment Report. He was also a member of the core writing team of the synthesis of the 2001 Third Assessment Report. He was elected as vice chairman of the IPCC in 1997 and in that position spearheaded the panel's activities on outreach and cross-cutting themes.

Pachauri's selection as IPCC chairman in 2002 was controversial. He succeeded Robert Watson, chief scientist of the World Bank and a former environmental official of the U.S. government under Bill Clinton, who was seeking a second five-year term. Watson had been a vocal supporter of a strong response to climate change and was backed by many environmental nongovernmental organizations. The U.S. government under President George W. Bush opposed Watson's reappointment and backed Pachauri, causing some to fear that this support and his connections to industry would undermine his ability to preserve the independence and integrity of the IPCC process.

Pachauri dispelled these notions quickly, however, standing up to various pressures and taking principled stands that have earned support from countries across the globe and the scientific community at large. Pachauri has a nuanced understanding of climate science and the myriad threats posed by climate change; he has proven adept at addressing the complex structure of the IPCC process. In addition, Pachauri's emphasis on outreach has been effective, increasing understanding and awareness of the IPCC's work in the media and governments around the world.

The IPCC's Fourth Assessment Report, which was produced in 2007 under Pachauri's leadership, is now widely perceived as a comprehensive and authoritative scientific assessment that makes clear that humans are changing the Earth's climate, primarily through the impact of greenhouse gas emissions, and that if left unchecked the impacts of climate change could be calamitous.

Norman Phillips (b. 1923)

Norman Phillips, a theoretical meteorologist, was the first person to successfully simulate the general circulation of the atmosphere using a numerical model of the type that came to be known as a general circulation model (GCM). Born in 1923, Phillips first became acquainted with weather prediction when he was assigned to the Azores as an Army Air Force weather forecaster during World War II. He went on to earn BS (1947) and PhD (1951) degrees in meteorology from the University of Chicago. Phillips joined the faculty at the Massachusetts Institute of Technology (MIT) in the 1950s, quickly advancing to chair the meteorology department. In 1974, he left MIT to serve as principal scientist at the U.S. National Weather Service, remaining there until his retirement in 1988.

In 1951, Phillips developed a two-layer model of the atmosphere considered to be the first that predicted changes in surface pressure. He expanded the model and then developed similar models for the global climate. By 1955, Phillips was working on a model using a cylindrical sheet of metal that was heated from the top and cooled from the bottom. This model produced a plausible representation of the large-scale movements of air masses that determine climate. His experiment was immediately hailed as a breakthrough, and Princeton University's Institute of Advanced Studies (where Phillips was conducting his research) quickly convened a conference to discuss it.

Despite its primitive nature, Phillips's model is now often regarded as the first true general circulation model. The model also laid important groundwork in the study of global warming, as current GCMs, based on Phillips's original, use transient climate simulations to project temperature changes under various scenarios. Among his many honors, in 2003 Phillips received the Benjamin Franklin Award for achievement in Earth science.

Roger R. Revelle (1909–1991)

Roger R. Revelle, a geochemist and oceanographer, spent his career conducting and supporting pioneering research on the interaction of the oceans and atmosphere. In his perhaps best-known discovery, Revelle and his colleague Hans Suess shattered a long-held scientific assumption when they published results demonstrating that the oceans would eventually stop absorbing all of the CO_2 being emitted by human activities—a critical discovery for understanding climate change.

Born in Seattle, Revelle attended Pomona College and then pursued graduate work in geology at the University of California, Berkeley. He joined the Scripps Institute of Oceanography as a research assistant in 1931, studying the chemical composition of sediment cores from the seafloor for his dissertation research and becoming an expert on the interaction of carbon and calcium compounds in seawater. After serving in the U.S. Navy during World War II, he returned to Scripps in 1946 and was named director in 1950.

One of Revelle's most important hires as director of Scripps was Hans Suess, a researcher at the University of Chicago who employed newly developed methods to measure the levels of carbon in tree rings. Together, Revelle and Suess began studying the carbon cycling of oceans, demonstrating how oceans might respond to increases in atmospheric CO_2 associated with the burning of fossil fuels. In 1957, they published an article demonstrating that the ocean was taking up only a fraction of the CO_2 emissions from human activities and describing the impacts of these emissions as a "large-scale geophysical experiment" (Revelle and Suess 1957).

The author of many important articles, Revelle also held significant administrative roles that helped to advance climate science and place climate on the political agenda. In the 1950s, he served as chair of the oceanography panel of the U.S. planning board for 1958's International Geophysical Year. In this capacity he made a crucial decision to support the establishment of monitoring stations to measure atmospheric CO_2. Charles David Keeling established these stations in Hawaii and Antarctica, and the data collected in those locations began to indicate that CO_2 levels were increasing. The stations continue to operate, and the data they collect plays a critical role in informing climate science and policy.

In addition, in 1965 Revelle served on the Presidential Science Advisory Committee Panel on Environmental Pollution and contributed to a report identifying carbon-induced warming as a potential environmental problem. A decade later, he chaired a National Academy of Sciences committee that issued a report warning that 40 percent of the CO_2 emitted from fossil fuel combustion had remained in the atmosphere and could result in a dangerous temperature increase of 6°C by the middle of the 21st century.

F. Sherwood Rowland (b. 1927)

F. Sherwood Rowland played a central role in the critical discovery that man-made CFCs threatened the Earth's ozone layer. For this work, Rowland and his colleague Mario Molina won the Nobel Prize for chemistry in 1995.

Born in Delaware, Ohio, Rowland moved quickly through his early education, graduating from high school and enrolling in Ohio Wesleyan University at the age of 16. At 18, with World War II continuing, he enlisted in the navy. After fulfilling a year of stateside service, Rowland returned to Ohio Wesleyan, completing his degree in 1948. He then earned an MS (1951) and a PhD in chemistry (1952) at the University of Chicago.

Rowland worked in the chemistry departments of Princeton University, the Brookhaven National Laboratory, and the University of Kansas before joining the faculty at the University of California, Irvine, in 1964. He pursued a variety of research topics, eventually taking an interest in the ultimate fate of CFCs. In 1973, Mario Molina joined the research group and, under Rowland's guidance, began studying this issue.

Rowland and Molina quickly realized that the fate of CFCs was not simply an academic question but one of grave environmental consequence. Their research has been hailed as identifying a critical environmental threat and the basis for developments that led to the landmark 1987 global environmental treaty, the Montreal Protocol on Substances that Deplete the Ozone Layer, as well as a series of equally important and precedent-setting amendments. Rowland received many honors for this path-breaking work, including the Tyler Prize (1983), the Japan Prize (1989), the Peter Debye Award (1993), the Roger Revelle Medal (1994), and the Nobel Prize for chemistry in 1995.

Rowland has also investigated the impact of methane gas on the atmosphere, demonstrating that methane is second only to CO_2 in contributing to global warming. In collaboration with Molina, Rowland's most recent work investigates the hydrocarbon and halocarbon composition of the atmosphere in remote locations and in polluted cities such as Santiago, Chile; Karachi, Pakistan; and Mexico City.

Stephen Schneider (b. 1945)

Stephen Schneider helped to pioneer the quantitative study of the climate system in the 1970s and became one of the most publicly engaged climate experts in the world. Since 1992, he has been a professor in the Department of Biological Sciences and senior fellow at the Center for Environment Science and Policy of the Institute for International Studies at Stanford University.

Born in New York, Schneider planned to become an astronomer but instead earned a BA in mechanical engineering and a PhD in mechanical engineering and plasma physics in 1971 at Columbia University. After the first Earth Day in 1970, Schneider became interested in advancing scientific understanding of the human impact on the environment, particularly the climate system. At the time, very little work was being done in the field, and new satellites and computers were making data and data analysis available for the first time. His research on the role of clouds in the energy balance was an important landmark in the study of climate feedback cycles. He was awarded a postdoctoral fellowship at the National Center for Atmospheric Research (NCAR) in 1972 and was a member of the scientific staff of NCAR from 1973 to 1996.

While at NCAR, Schneider conducted research on aerosols and greenhouse gases and quickly became a public figure when he proposed that aerosols were cooling the planet while greenhouse gases were warming it (Nuzzo 2005). Many scientists at the time disapproved of Schneider's public engagement because they thought it undermined the integrity of the science. Schneider went on to become an expert in the communication of complex ideas, authoring both scientific and general interest books and articles and appearing on radio and television. In 1992, the MacArthur Foundation awarded him a "genius grant" in recognition of his extraordinary ability to integrate disciplines and communicate the complexities of climate change to a wide variety of audiences.

Susan Solomon (b. 1956)

One of the world's leading atmospheric scientists, Susan Solomon served as the cochair of the IPCC's Working Group I, which explores the physical science basis of climate change, during the writing of its 2007 Fourth Assessment Report. Earlier in her career Solomon played an important role in the research that proved a link between CFCs and the depletion of the ozone layer.

Born in Chicago, Solomon became interested in science by watching Jacques Cousteau's expeditions on television. She attended the Illinois Institute of Technology (BS 1977) and received a PhD in atmospheric chemistry from the University of California at Berkeley in 1981. She continued her work at the National Center for Atmospheric Research in Boulder, Colorado, focusing on reactions in the mesosphere, thermosphere and stratosphere. Solomon is currently the head of the Chemistry and Climate Processes Group of the National Oceanic and Atmospheric Administration Chemical Sciences Division.

In 1985, British scientists announced research showing that there was a hole in the ozone layer above Antarctica and serious depletion elsewhere. Building on the work of Rowland and Molina, Solomon and colleagues proposed that the rapid depletion of ozone over the Antarctic was caused by reactions between man-made CFCs and by ice crystals in clouds over the Antarctic, which accelerated the reactions (Solomon 1986). In 1986 and 1987, she helped lead research expeditions to the area, gathering evidence that provided the first clear proof that CFCs were indeed destroying stratospheric ozone. Solomon's discoveries assisted efforts to significantly strengthen the 1987 Montreal Protocol on Substances that Deplete the Ozone Layer and helped to create a new field of atmospheric chemistry examining gas-particle interactions.

Among her many honors, Solomon has received the National Medal of Science in 1994 and the Blue Planet Prize in 2004.

Nicholas Stern (b. 1946)

Nicholas Stern directed a comprehensive review of the costs and benefits of climate change and the economics of stabilizing greenhouse gases in the atmosphere. Published in 2006, *The Economics of Climate Change: The Stern Review* argued that the overall costs to the world economy of unchecked climate change would far outweigh the costs of reducing greenhouse gas emissions and that governments need to take strong action in the immediate future (Stern 2006). Stern's reputation as a leading economist was critical to establishing the document's credibility.

Born in London, Stern earned his PhD in economics at Nuffield College, Oxford and served as a lecturer at Cambridge University from 1970 to 1977 and as a professor of economics at the University of Warwick from 1978 to 1987. He taught at the London School of Economics from 1986 until 1993, when he became chief economist at the European Bank for Reconstruction and Development. He then served as chief economist at the World Bank from 2000 to 2003.

In 2003, Stern was appointed head of the UK Government Economic Service. In that role he undertook two large-scale economic reviews. The first was the Commission for Africa; the commission's report, published in 2005, formed the background for that year's Gleneagles G8 Summit.

In 2006, Stern began work on the second of these reviews—a report on the economic implications of climate change. Though Stern began the work as a self-described "climate agnostic," he quickly became convinced that decisive action was needed to avoid serious economic, social, and environmental costs. Indeed, according to the Stern report, following a business-as-usual scenario of greenhouse gas emissions would result in a 20 percent reduction in per capita consumption over the coming century as climate change impacts reduce the availability of resources. Accordingly, the Stern report recommends an investment of 1 percent of global GDP per year in order to mitigate the effects of climate change.

Stern left his position with the Government Economic Service in 2007 to return to the London School of Economics. He currently serves as the IG Patel Professor of Economics and Government and heads the new India Observatory within the Asia Research Center.

Hans E. Suess (1909–1993)

Born in Vienna in 1909, Hans E. Suess was part of a dynasty of famous Austrian scientists and grew up in an environment of scientific excellence. Preceded at the University of Vienna by his father, Franz Eduard Suess, and grandfather, Eduard Suess, Hans completed his PhD in physical chemistry in 1936 before taking a position at the Swiss Technical University in Zurich and later at the Institute for Physical Chemistry in Hamburg, Germany.

During World War II, Suess was part of a team of German scientists studying atomic energy and an advisor to the production of heavy water (used in production of nuclear materials) in a Norwegian plant in Vermok. After the war, Suess worked on a shell model of the atomic nucleus with future Nobel Prize winner Hans Jensen.

In 1950, Suess joined the University of Chicago's Institute for Nuclear Studies (now the Fermi Institute). He conducted research in the field of cosmochemistry, investigating the abundance of certain elements in meteorites with Harold Urey. In 1955, Suess found that tree rings, and by inference the atmosphere, were being contaminated by stable carbon that could only have come from the burning of fossil fuels (Suess 1955). Shortly thereafter, Roger Revelle recruited Suess to join a small group of geochemists and geophysicists at the Scripps Institute of Oceanography, which later joined the University of California, San Diego (UCSD).

At the time, a group of scientists was exploring the distribution of CO_2 among the major terrestrial reservoirs—the atmosphere, land vegetation, and the ocean. Revelle's and Suess's 1957 paper was the first to call attention to the fact that the growing quantity of CO_2 in the atmosphere could cause global warming over time.

Suess's interests continued to widen, and he published in a number of fields, including meteorology and carbon dating. He taught at Scripps and in the Department of Chemistry at UCSD until the end of his scientific career. He was elected to the National Academy of Sciences in 1966 and received numerous other honors.

John Tyndall (1820–1893)

In discovering that some atmospheric gases absorb far more heat than others, John Tyndall helped to build our modern under-standing of the greenhouse effect and climate change. The son of a policeman, Tyndall was born in Carlow, Ireland. He left school at 17 to work as a draftsman and civil engineer. He was dis-missed from the English Ordnance Survey for protesting unfair treatment of Irish employees and in 1847 became a teacher of mathematics, surveying, and engineering physics at the Quaker School at Queenswood College in Hampshire, England.

Urged by colleagues to continue his education, Tyndall went abroad to the University of Marburg in Germany in 1851, completing coursework for a PhD in physics in two years. He re-turned to England and gave several popular lecture series at the Royal Institution of Great Britain, which hired him as a professor of natural philosophy in 1853. Tyndall was a skilled and an en-thusiastic lecturer on a wide range of scientific topics and became an important figure in the popularization of science.

In 1859, he began testing various gases to see how they re-spond to infrared radiation and found that some atmospheric gases, such as carbon dioxide and water vapor, almost com-pletely block the transmission of heat. Tyndall recognized that this phenomenon played a critical role in maintaining a surface temperature hospitable to life.

Stephen Zebiak (b. 1956)

Stephen Zebiak is director general of the International Research Institute for Climate and Society and has been a leading expert in ocean-atmosphere interaction, climate variability, and climate pre-diction for 20 years. In the 1980s, Zebiak and Mark Cane created the first dynamical model to successfully predict the El Niño Southern Oscillation (Cane, Zebiak, and Dolan 1986; Cane and Zebiak 1987; Zebiak and Cane 1987). Since then, he has contributed to hundreds of articles, papers, and reports that have significantly advanced our understanding of El Niño, how to predict it, its impacts in different parts of the world, and how to prepare for these impacts.

Born in Skowhegan, Maine, Zebiak received his undergraduate degree in mathematics from the Massachusetts Institute of Technology in 1978, an MS in applied mathematics from Rensselaer Polytechnic Institute in 1979, and a PhD in meteorology from MIT in 1984. After leaving MIT, he joined the Lamont-Doherty Earth Observatory at Columbia University as a postdoctoral researcher. In 1993, he received an appointment as senior research scientist at both LDEO and IRI. In 2003, Zebiak assumed leadership of IRI when he was named director general. A unit of Columbia University, IRI uses science-based approaches to enhance society's ability to understand, anticipate, and manage climate variability, climate change, and climate risk in order to improve human welfare.

Zebiak has received a variety of honors and served on numerous important national and international advisory and scientific committees, including the Atlantic Climate Change Program, the Pan American Climate Studies Program, the American Meteorological Society Committee on Climate Variations, and the Cooperative Institute for Climate Applications and Research. He has also worked with the International CLIVAR Working Groups, the U.S. CLIVAR Seasonal-to-Interannual Modeling and Prediction Panel, the Asia-Pacific Economic Community Climate Network, and the Committee on Strategic Guidance for National Science Foundation Support of Research in the Atmospheric Sciences.

References

Bolin, B., and E. Eriksson. 1959. "Changes in the Carbon Dioxide Content of the Atmosphere and Sea Due to Fossil Fuel Combustion." In *The Atmosphere and the Sea in Motion*, edited by Bert Bolin, 130–142. New York: Rockefeller Institute Press.

Broecker, W. S. 1966. "Absolute Dating and the Astronomical Theory of Glaciation." *Science* 151:299–304.

Broecker, W. S., and J. van Donk. 1970. "Insolation Changes, Ice Volumes, and the O^{18} Record in Deep-Sea Cores." *Reviews of Geophysics and Space Physics* 8:169–178.

Broecker, W. S., D. M. Peteet, and D. Rind. 1985. "Does the Ocean-Atmosphere System Have More than One Stable Mode of Operation?" *Nature* 315:21–26.

Callendar, G. S. 1938. "The Artificial Production of Carbon Dioxide and Its Influence on Temperature." *Quarterly Journal of the Royal Meteorological Society* 64:223–240.

Cane, M., S. E. Zebiak, and S. C. Dolan. 1986. "Experimental Forecasts of El Nino." *Nature* 321:827–832.

Cane, M., and S. E. Zebiak. 1987. "Prediction of El Niño Events Using a Physical Model." In *Atmospheric and Oceanic Variability*, edited by H. Cattle. Reading, UK: Royal Meteorological Society Press.

Chen, D., S. E. Zebiak, A. J. Busalacchi, and M. A. Cane. 1995. "An Improved Procedure for El Niño Forecasting: Implications for Predictability." *Science* 269:1699–1702.

Fourier, J. 2004 (1878). *The Analytical Theory of Heat*, trans. A. Freeman. New York: Dover.

Gore, Al. 1992. *Earth in the Balance: Ecology and the Human Spirit*. Boston: Houghton Mifflin.

Gore, Al. 2006. *An Inconvenient Truth: The Planetary Emergency of Global Warming and What We Can Do about It*. New York: Rodale.

Hansen, J., D. Johnson, A. Lacis, S. Lebedeff, P. Lee, D. Rind, and G. Russell. 1981. "Climate Impact of Increasing Atmospheric Carbon Dioxide." *Science* 213:957–966.

Mann, M., R. S. Bradley, and M. K. Hughes. 1998. "Global-Scale Temperature Patterns and Climate Forcings over the Past Six Centuries." *Nature* 392:779–787.

Mann, M., R. S. Bradley, and M. K. Hughes. 1999. "Northern Hemisphere Temperatures during the Last Millennium: Inferences, Uncertainties and Limitations." *Geophysical Research Letters* 26:759–762.

Nuzzo, R. 2005. "Profile of Stephen H. Schneider." *Proceedings of the National Academy of Sciences* 102 (44): 15725–15727.

Phillips, N. A. 1956. "The General Circulation of the Atmosphere: A Numerical Experiment." *Quarterly Journal of the Royal Meteorological Society* 82:123–164.

Revelle, R., and H. E. Suess. 1957. "Carbon Dioxide Exchange between Atmosphere and Oceans and the Question of an Increase of Atmospheric CO_2 during the Past Decade." *Tellus* 9:18–27.

Rowland, F. S., and M. Molina. 1974. "Stratospheric Sink for Chlorofluoromethanes: Chlorine-Atom Catalysed Destruction of Ozone." *Nature* 249 (5460): 810–812.

Solomon, S., R. R. Garcia, F. S. Rowland, and D. J. Wuebbles. 1986. "On the Depletion of Antarctic Ozone." *Nature* 321:755–758.

Stern, Nicholas. 2006. *The Economics of Climate Change: The Stern Review.* Cambridge, UK: Cambridge University Press.

Suess, Hans E. 1955. "Radiocarbon Concentration in Modern Wood." *Science* 122:415–417.

U.S. Congress, House of Representatives, Committee on Oversight and Government Reform. 2007a. "Political Interference with Government Climate Change Science: Testimony of James Hansen." [Online testimony; retrieved 7/18/08.] http://oversight.house.gov/documents/20070319105800–43018.pdf.

U.S. Congress, House of Representatives, Committee on Oversight and Government Reform. 2007b. "Committee Report: White House Engaged in Systematic Effort to Manipulate Climate Change Science." [Online report; retrieved 7/18/08.] http://oversight.house.gov/story.asp?ID=1653.

Zebiak, S. E., and M. A. Cane. 1987. "A Model for El Niño–Southern Oscillation." *Monthly Weather Review* 115:2262–2278.

6

Data and Documents

This chapter contains data, diagrams, and excerpts from important documents relating to climate change science and policy.

Like glass, gases in the atmosphere let in light yet prevent heat from escaping. This natural warming of the planet as a result of heat-trapping gases is called the greenhouse effect. The mechanisms involved in the greenhouse effect are displayed in Figure 6.1.

Figure 6.2, taken from the Intergovernmental Panel on Climate Change's (IPCC) Fourth Assessment Report, shows the relationship between incoming and outgoing radiation in the Earth system. The Earth's temperature is in part the result of the interplay between certain elements in the radiation balance. Disruptions to the Earth's radiation balance can occur when (1) incoming solar radiation changes, (2) the fraction of solar radiation that is reflected changes, or (3) the longwave radiation from the Earth back toward space is altered. Over long periods of time and across the whole world, disruptions in the radiation balance may lead to the protracted heating or cooling of the climate system.

Radiative forcing occurs when something causes the radiation balance of the Earth's system to change, contributing to long-term warming or cooling of the planet. This forcing may occur as a result of perturbations in the system or the introduction of foreign agents, including anthropogenic greenhouse gases. As Figure 6.3 illustrates, both natural and man-made phenomena contribute to changes in the Earth's radiation balance.

Table 6.1 lists greenhouse gases and their global warming potential. The global warming potential (GWP) of a substance is

FIGURE 6.1
The Greenhouse Effect

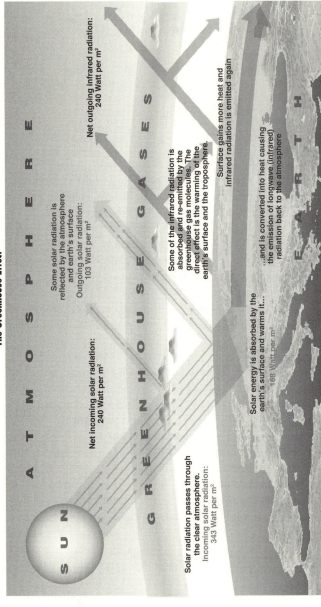

S U N

A T M O S P H E R E

G R E E N H O U S E   G A S E S

E A R T H

Solar radiation passes through the clear atmosphere.
Incoming solar radiation:
343 Watt per m²

Net incoming solar radiation:
240 Watt per m²

Some solar radiation is reflected by the atmosphere and earth's surface
Outgoing solar radiation:
103 Watt per m²

Net outgoing infrared radiation:
240 Watt per m²

Some of the infrared radiation is absorbed and re-emitted by the greenhouse gas molecules. The direct effect is the warming of the earth's surface and the troposphere.

Surface gains more heat and infrared radiation is emitted again

Solar energy is absorbed by the earth's surface and warms it...
168 Watt per m²

...and is converted into heat causing the emission of longwave (infrared) radiation back to the atmosphere

FIGURE 6.2
Radiation Balance

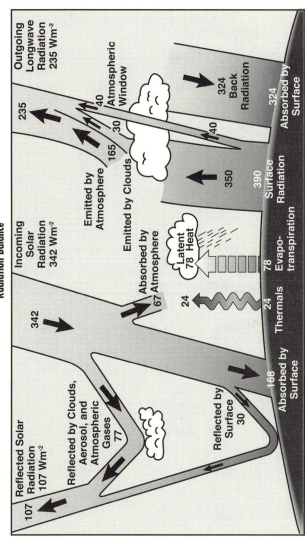

Source: Reprinted with permission from Le Treut, H., R. Somerville, U. Cubasch, Y. Ding, C. Mauritzen, A. Mokssit, T. Peterson, and M. Prather. 2007. "Historical Overview of Climate Change." In *Climate Change 2007: The Physical Science Basis. Contribution of Working Group I to the Fourth Assessment Report of the Intergovernmental Panel on Climate Change,* edited by S. Solomon, D. Qin, M. Manning, Z. Chen, M. Marquis, K. B. Averyt, M. Tignor, and H. L. Miller. Cambridge, UK: Cambridge University Press.

FIGURE 6.3
Components of Radiative Forcing

Components of radiative forcing for principal emissions

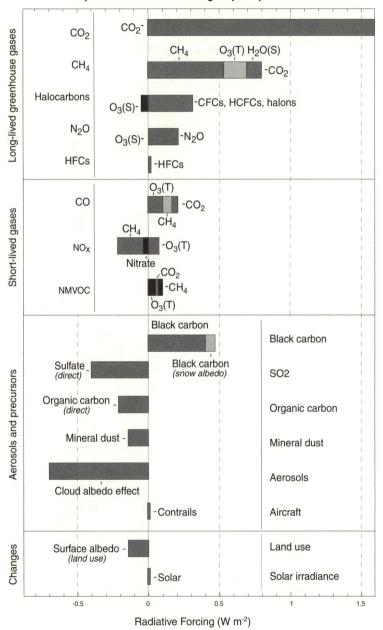

Radiative Forcing (W m⁻²)

Source: Reprinted with permission from Forster, P., V. Ramaswamy, P. Artaxo, T. Berntsen, R. Betts, D. W. Fahey, J. Haywood, J. Lean, D. C. Lowe, G. Myhre, J. Nganga, R. Prinn, G. Raga, M. Schulz, and R. Van Dorland. 2007. "Changes in Atmospheric Constituents and in Radiative Forcing." In *Climate Change 2007: The Physical Science Basis. Contribution of Working Group I to the Fourth Assessment Report of the Intergovernmental Panel on Climate Change,* edited by S. Solomon, D. Qin, M. Manning, Z. Chen, M. Marquis, K. B. Averyt, M.Tignor, and H. L. Miller. Cambridge, UK: Cambridge University Press.

TABLE 6.1

Complete List of Greenhouse Gases and Their Global Warming Potentials

Industrial Designation or Common Name	Chemical Formula	Lifetime (years)	Radiative Efficiency	Global Warming Potential for a Given Time Horizon			
				SAR (100-yr)	20-yr	100-yr	500-yr
Carbon dioxide	CO_2	See below a	b1.4×10^{-5}	1	1	1	1
Methane	CH_4	12c	3.7×10^{-4}	21	72	25	8
Nitrous oxide	N_2O	114	3.03×10^{-3}	310	289	298	153
Substances Controlled by the Montreal Protocol							
CFC-11	CCl_3F	45	0.25	3,800	6,730	4,750	1,620
CFC-12	CCl_2F_2	100	0.32	8,100	11,000	10,900	5,200
CFC-13	$CClF_3$	640	0.25		10,800	14,400	16,400
CFC-113	CCl_2FCClF_2	85	0.30	4,800	6,540	6,130	2,700
CFC-114	$CClF_2CClF_2$	300	0.31		8,040	10,000	8,730
CFC-115	$CClF_2CF_3$	1,700	0.18		5,310	7,370	9,990
Halon-1301	$CBrF_3$	65	0.32	5,400	8,480	7,140	2,760
Halon-1211	$CBrClF_2$	16	0.30		4,750	1,890	575
Halon-2402	$CBrF_2CBrF_2$	20	0.33		3,680	1,640	503
Carbon tetrachloride	CCl_4	26	0.13	1,400	2,700	1,400	435
Methyl bromide	CH_3Br	0.70	0.01		17	5	1
Methyl chloroform	CH_3CCl_3	5.00	0.06		506	146	45
HCFC-22	$CHClF_2$	12.00	0.20	1,500	5,160	1,810	549
HCFC-123	$CHCl_2CF_3$	1.30	0.14	90	273	77	24

(continues)

TABLE 6.1 (continued)
Complete List of Greenhouse Gases and Their Global Warming Potentials

Industrial Designation or Common Name	Chemical Formula	Lifetime (years)	Radiative Efficiency	Global Warming Potential for a Given Time Horizon			
				SAR (100-yr)	20-yr	100-yr	500-yr
HCFC-124	$CHClFCF_3$	5.80	0.22	470	2,070	609	185
HCFC-141b	CH_3CCl_2F	9.30	0.14		2,250	725	220
HCFC-142b	CH_3CClF_2	17.90	0.20	1,800	5,490	2,310	705
HCFC-225ca	$CHCl_2CF_2CF_3$	1.90	0.20		429	122	37
HCFC-225cb	$CHClFCF_2ClF_2$	5.80	0.32		2,030	595	181
Hydrofluorocarbons							
HFC-23	CHF_3	270	0.19	11,700	12,000	14,800	12,200
HFC-32	CH_2F_2	4.90	0.11	650	2,330	675	205
HFC-125	CHF_2CF_3	29	0.23	2,800	6,350	3,500	1,100
HFC-134a	CH_2FCF_3	14	0.16	1,300	3,830	1,430	435
HFC-143a	CH_3CF_3	52	0.13	3,800	5,890	4,470	1,590
HFC-152a	CH_3CHF_2	1.40	0.09	140	437	124	38
HFC-227ea	CF_3CHFCF_3	34.20	0.26	2,900	5,310	3,220	1,040
HFC-236fa	$CF_3CH_2CF_3$	240	0.28	6,300	8,100	9,810	7,660
HFC-245fa	$CHF_2CH_2CF_3$	7.60	0.28		3,380	1,030	314
HFC-365mfc	$CH_3CF_2CH_2CF_3$	8.60	0.21		2,520	794	241
HFC-43-10mee	$CF_3CHFCHFCF_2CF_3$	15.90	0.40	1,300	4,140	1,640	500
Perfluorinated compounds							
Sulphur hexafluoride	SF_6	3,200	0.52	23,900	16,300	22,800	32,600

Nitrogen trifluoride	NF_3	740		0.21	12,300	17,200	20,700
PFC-14	CF_4	50,000	6,500	0.10	5,210	7,390	11,200
PFC-116	C_2F_6	10,000	9,200	0.26	8,630	12,200	18,200
PFC-218	C_3F_8	2,600	7,000	0.26	6,310	8,830	12,500
PFC-318	$c\text{-}C_4F_8$	3,200	8,700	0.32	7,310	10,300	14,700
PFC-3-1-10	C_4F_{10}	2,600	7,000	0.33	6,330	8,860	12,500
PFC-4-1-12	C_5F_{12}	4,100		0.41	6,510	9,160	13,300
PFC-5-1-14	C_6F_{14}	3,200	7,400	0.49	6,600	9,300	13,300
PFC-9-1-18	$C_{10}F_{18}$	>1,000		0.56	>5,500	>7,500	>9,500
Trifluoromethyl sulphur pentafluoride	SF_5CF_3	800		0.57	13,200	17,700	21,200
Fluorinated ethers							
HFE-125	CHF_2OCF_3	136		0.44	13,800	14,900	8,490
HFE-134	CHF_2OCHF_2	26		0.45	12,200	6,320	1,960
HFE-143a	CH_3OCF_3	4.30		0.27	2,630	756	230
HCFE-235da2	$CHF_2OCHClCF_3$	2.60		0.38	1,230	350	106
HFE-245cb2	$CH_3OCF_2CHF_2$	5.10		0.32	2,440	708	215
HFE-245fa2	$CHF_2OCH_2CF_3$	4.90		0.31	2,280	659	200
HFE-254cb2	$CH_3OCF_2CHF_2$	2.60		0.28	1,260	359	109
HFE-347mcc3	$CH_3OCF_2CF_2CF_3$	5.20		0.34	1,980	575	175
HFE-347pcf2	$CHF_2CF_2OCH_2CF_3$	7.10		0.25	1,900	580	175
HFE-356pcc3	$CH_3OCF_2CF_2CHF_2$	0.33		0.93	386	110	33
HFE-449sl (HFE-7100)	$C_4F_9OCH_3$	3.80		0.31	1,040	297	90
HFE-569sf2 (HFE-7200)	$C_4F_9OC_2H_5$	0.77		0.30	207	59	18
HFE-43-10pccc124 (H-Galden 1040x)	$CHF_2OCF_2OC_2F_4OCHF_2$	6.30		1.37	6,320	1,870	569
HFE-236ca12 (HG-10)	$CHF_2OCF_2OCHF_2$	12.10		0.66	8,000	2,800	860
HFE-338pcc13 (HG-01)	$CHF_2OCF_2CF_2OCHF_2$	6.20		0.87	5,100	1,500	460

(continues)

TABLE 6.1 (continued)
Complete List of Greenhouse Gases and Their Global Warming Potentials

Industrial Designation or Common Name	Chemical Formula	Lifetime (years)	Radiative Efficiency	Global Warming Potential for a Given Time Horizon			
				SAR (100-yr)	20-yr	100-yr	500-yr
Perfluoropolyethers							
PFPMIE	$CF_3OCF(CF_3)CF_2OCF_2OCF_3$	800	0.65		7,620	10,300	12,400
Hydrocarbons and other compounds – Direct Effects							
Dimethylether	CH_3OCH_3	0.02	0.02		1	1	<<1
Methylene chloride	CH_2Cl_2	0.38	0.03		31	9	3
Methyl chloride	CH_3Cl	1	0.01		45	13	4

Source: Forster, P., V. Ramaswamy, P. Artaxo, T. Berntsen, R. Betts, D. W. Fahey, J. Haywood, J. Lean, D. C. Lowe, G. Myhre, J. Nganga, R. Prinn, G. Raga, M. Schulz, and R. Van Dorland. 2007. "Changes in Atmospheric Constituents and in Radiative Forcing." In *Climate Change 2007: The Physical Science Basis. Contribution of Working Group I to the Fourth Assessment Report of the Intergovernmental Panel on Climate Change*, edited by S. Solomon, D. Qin, M. Manning, Z. Chen, M. Marquis, K. B. Averyt, M. Tignor, and H. L. Miller. Cambridge, UK: Cambridge University Press.

FIGURE 6.4
The Keeling Curve: CO_2 Concentrations at Mauna Loa Observatory, Hawaii

Monthly Carbon Dioxide Concentration
parts per million

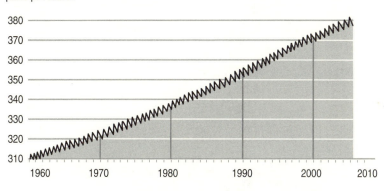

Source: Reprinted with permission from Scripps Institute of Oceanography. 2008. "Welcome to Scripps CO_2."
[Online information; retrieved 7/22/08.] http://scrippsco2.ucsd.edu/home/index.php

the cumulative warming effect that a single unit of the substance will have on the climate system over a certain period of time. To create a reference standard, scientists always assign the value of 1 as the GWP of a unit of CO_2. When the expected lifetime and radiative efficiency are taken into account, the global warming potential of a gas is a measure of its contribution to global warming.

The graph shown in Figure 6.4, commonly known as the Keeling Curve, depicts CO_2 atmospheric concentrations collected in Mauna Loa, Hawaii, beginning in 1958. This work was initiated by Charles David Keeling, formerly of the Scripps Institution of Oceanography, who was the first person to take regular measurements of atmospheric carbon dioxide concentrations.

The graph shows a steady increase in atmospheric CO_2 concentrations as the result of increasing use of fossil fuels. The Keeling Curve also shows a cyclic variation of about 5 parts per million (ppm) by volume in each year. This variation represents the seasonal change in uptake of CO_2 by the world's land vegetation.

As indicated by the map in Figure 6.5, the United States was the single largest emitter of greenhouse gases in the world in 2000, although China's emissions are believed to have surpassed

FIGURE 6.5
Total CO$_2$ Emissions by Country in 2000

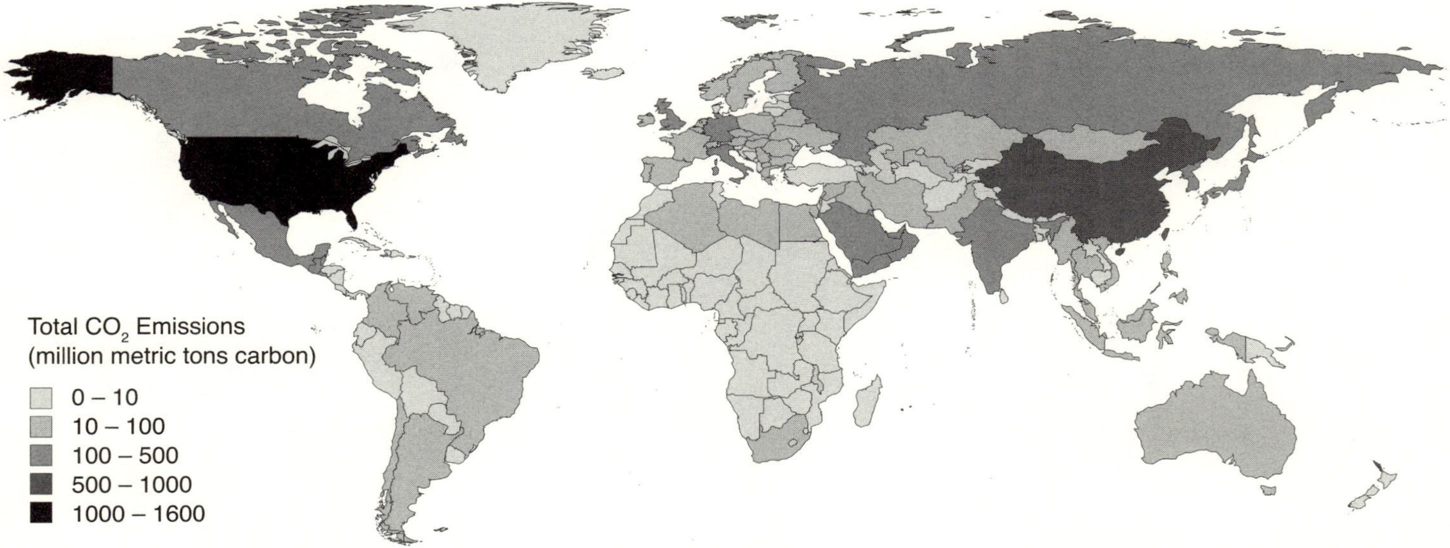

Total CO$_2$ Emissions
(million metric tons carbon)

- 0 – 10
- 10 – 100
- 100 – 500
- 500 – 1000
- 1000 – 1600

Source: Marland, G., T. A. Boden, and R. J. Andres. "Global Regional and National Fossil Fuel CO$_2$ Emissions." In *Trends: A Compendium of Data on Global Climate Change.* Oak Ridge, TN: Carbon Dioxide Information Analysis Center, Oak Ridge National Laboratory, U.S. Department of Energy. Maps produced by the Center for Sustainability and the Global Environment. Reprinted with permission from the *Annual Review of Public Health,* vol. 29. Copyright 2008 by Annual Reviews. www.annualreviews.org.

FIGURE 6.6a
Per Capita Greenhouse Gas Emissions Growth in Non-Annex I Countries

Per Capita Emissions:
Per Capita Emissions in 2003 (Annex I and non-Annex I)
Growth in Per Capita Emissions since 1990 (non-Annex I)

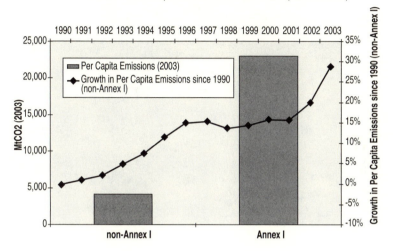

the U.S. emissions in 2007 or 2008. Emissions in other developing countries, including India and Brazil, continue to rise as well, while the emissions levels of many other developing countries, particularly in Africa, are still quite low.

Figures 6.6a and 6.6b show per capita emissions in 2003 for both industrialized and developing countries. The line in Figure 6.6a shows the growth in per capita emissions for developing country, or non-Annex I, countries since 1990, while the line in Figure 6.6b represents the growth in per capita emissions since 1990 for Annex I, or industrialized, countries. While industrialized countries emit considerably more per capita, individual and collective developing country emissions are increasing rapidly.

By overlaying global average temperatures and carbon dioxide data, the graph in Figure 6.7 provides clear evidence for a link between rising temperatures and carbon dioxide.

Extending the time scale to 400,000 years ago, the graph in Figure 6.8 presents more evidence for a link between atmospheric carbon dioxide and global average temperature.

FIGURE 6.6b
Per Capita Greenhouse Gas Emissions Growth in Annex I Countries

Per Capita Emissions:
Per Capita Emissions in 2003 (Annex I and non-Annex I)
Growth in Per Capita Emissions since 1990 (Annex I)

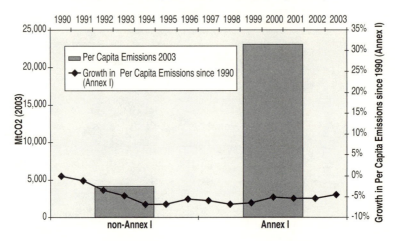

Source: Reprinted with permission from World Resources Institute. 2003. "Climate Analysis Indicators Tool." [Online information; retrieved 7/22/08.] http://cait.wri.org/

As a result of agricultural improvements and increased access to health care, the 20th century was marked by dramatic growth in world population. Increases in population have been matched by rising CO_2 emissions, which have, in turn, contributed to increased surface air temperatures.

Though the growth rate has slowed, world population is expected to continue to increase, reaching 9 billion by the mid-21st century. This is shown in Figure 6.10.

The chart shown in Figure 6.11, developed by the IPCC for its Fourth Assessment Report, makes explicit the potential effects of projected temperature increases on human and natural systems. Solid lines link impacts, while dotted arrows indicate effects that continue with increasing temperature. Text placement corresponds to the approximate onset, with respect to temperature, of a given impact. Adaptation is not included in the assessment.

Table 6.2, also developed by the IPCC for its Fourth Assessment Report, provides more information on the potential effects of climate change.

FIGURE 6.7
Global Average Temperatures and Carbon Dioxide Concentrations, 1880–2004

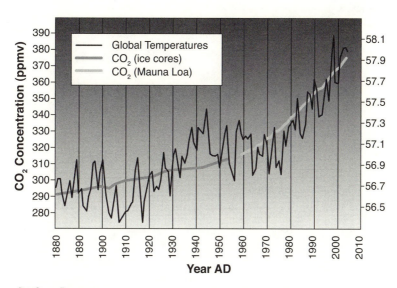

Data Source, Temperature:
  ftp://ftp.ncdc.noaa.gov/pub/data/anomalies/annual.land_and_ocean.90S.90N.df_1901-2002 mean.dat
Data Source CO_2 (Siple Ice Cores):
  http://cdiac.esd.ornl.gov/ftp/trends/co2/siple2.013
Date Source CO_2 (Mauna Loa):
  http://cdiac.esd.ornl.gov/ftp/trends/co2/maunaloa.co2
  http://esri.noaa.gov.gmd.webdata/ccgg/trends/co2_mm_mlo.date
Graphic Design: Michael Ernst, Woods Hole Research Center:
  http://www.whrc.org/resources/online_publications/warming_earth/images/Fig2-CO2-Temp.jpg
  Reprinted with permission.

As Figure 6.12 illustrates, climate change will increase the vulnerability of those whose lives and livelihoods are already threatened by extreme poverty, disease, extreme weather, and climate variability.

Figure 6.13 displays the level of mitigation expected from various sectors around the world as a function of the price of carbon. In some sectors, including agriculture, an increase in the price of carbon will have dramatic effects on the level of mitigation. In other sectors, mitigation efforts will remain roughly equal

FIGURE 6.8

Temperature and CO₂ Concentration in the Atmosphere over the Past 400,000 Years (from the Vostok Ice Core)

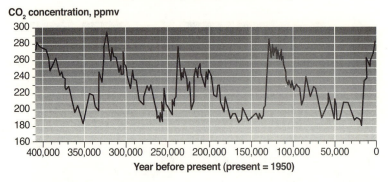

CO₂ concentration, ppmv

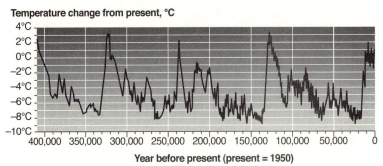

Temperature change from present, °C

Source: Reprinted by permission from MacMillan Publishers: Petit, J. R., J. Jouzel, et al. 1999. "Climate and Atmospheric History of the Past 420,000 Years from the Vostok Ice Core in Antarctica," *Nature* 399 (3 June), 429–436. This graph is also available from the UNEP/GRID-Arendal Maps and Graphics Library at http://maps.grida.no/go/graphic/temperature-and-co2-concentration-in-the-atmosphere-over-the-past-400-000-years.

regardless of whether carbon is priced closer to $20 per gigaton (Gt) or $100 per Gt.

Figures 6.14a and 6.14b present estimates of global economic mitigation potential in 2030, represented on the horizontal axes. Economic mitigation potential takes into account social costs and benefits of mitigation activities and assumes that market efficiency is improved by policies that remove barriers to mitigation.

These estimates are based on both bottom-up and top-down studies. Bottom-up approaches are based on assessments of mitigation options, emphasizing technologies and regulation. They

FIGURE 6.9
Population, CO$_2$ Emissions, and Temperature, 1950–2001

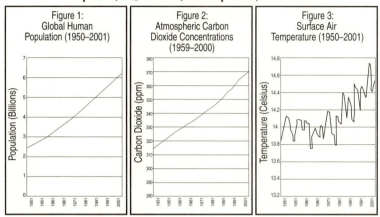

Source: Reprinted with permission from Meyerson, F. A. B. 2004. "Executive Summary: Population Dynamics and Global Climate Change." [Online information; retrieved 7/22/08.] http://www.prcdc.org/files/2002%20Population %20Dynamics%20and%20Global%20Climate%20Change.
Data Sources: 1) United Nations Department for Economic and Social Information and Policy Analysis, Population Division. 2000. *World Population Prospects: The 2000 Revision.* New York: United Nations; 2) Scripps Institution of Oceanography; 3) Goddard Institute for Space Studies.

FIGURE 6.10
World Population, 1950–2050

Population (billions)

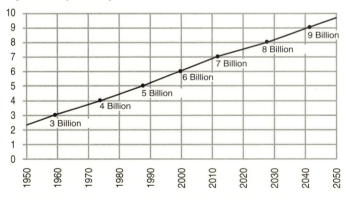

Source: U.S. Census Bureau International Database, July 2007. Available at: http://www.census.gov/ipc/ www/img/worldpop.gif

FIGURE 6.11
Projected Effects of Climate Change and Their Likelihood

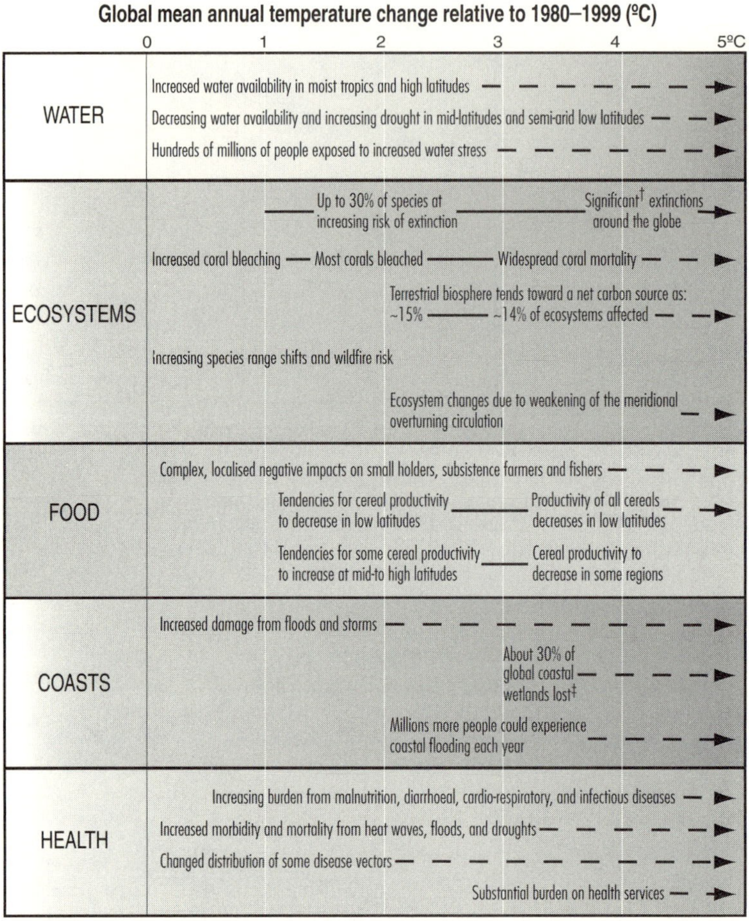

Global mean annual temperature change relative to 1980–1999 (°C)

	0	1	2	3	4	5°C

WATER
Increased water availability in moist tropics and high latitudes
Decreasing water availability and increasing drought in mid-latitudes and semi-arid low latitudes
Hundreds of millions of people exposed to increased water stress

ECOSYSTEMS
Up to 30% of species at increasing risk of extinction
Significant[†] extinctions around the globe
Increased coral bleaching — Most corals bleached — Widespread coral mortality
Terrestrial biosphere tends toward a net carbon source as: ~15% — ~14% of ecosystems affected
Increasing species range shifts and wildfire risk
Ecosystem changes due to weakening of the meridional overturning circulation

FOOD
Complex, localised negative impacts on small holders, subsistence farmers and fishers
Tendencies for cereal productivity to decrease in low latitudes
Productivity of all cereals decreases in low latitudes
Tendencies for some cereal productivity to increase at mid-to high latitudes
Cereal productivity to decrease in some regions

COASTS
Increased damage from floods and storms
About 30% of global coastal wetlands lost[‡]
Millions more people could experience coastal flooding each year

HEALTH
Increasing burden from malnutrition, diarrhoeal, cardio-respiratory, and infectious diseases
Increased morbidity and mortality from heat waves, floods, and droughts
Changed distribution of some disease vectors
Substantial burden on health services

† Significant is defined here as more than 40%   ‡ Based on average rate of sea level rise of 4.2 mm/year

Source: Reprinted with permission from Intergovernmental Panel on Climate Change. 2007. "Summary for Policymakers." In *Climate Change 2007: Impacts, Adaptation and Vulnerability. Contribution of Working Group II to the Fourth Assessment Report of the Intergovernmental Panel on Climate Change,* edited by M. L. Parry, O. F. Canziani, J. P. Palutikof, P. J. van der Linden, and C. E. Hanson. Cambridge, UK: Cambridge University Press.

TABLE 6.2

More IPCC Information on Projected Impacts and Their Likelihood

Phenomenon and Direction of Trend	Likelihood of Future Trends Based on Projections for 21st Century Using SRES Scenarios	Agriculture, Forestry, and Ecosystems	Water Resources	Human Health	Industry, Settlement, and Society
Over most land areas, warmer and fewer cold days and nights, warmer and more frequent hot days and nights	Virtually certain	Increased yields in colder environments; decreased yields in warmer environments; increased insect outbreaks	Effects on water resources relying on snow melt, effects on some water supply	Reduced human mortality from decreased cold exposure	Reduced energy demand for heating; increased demand for cooling; declining air quality in cities; reduced disruption to transport due to snow, ice; effects on winter tourism
Warm spells/heat waves increase in frequency over most land areas	Very likely	Reduced yields in regions due to heat stress; wildfire danger increase	Increased water demand; water quality problems (e.g., algae blooms)	Increased risk of heat-related mortality, especially for the elderly, chronically sick, very young, and socially isolated	Reduction in the quality of life for people in warm areas without appropriate housing; impacts on elderly, very young, and poor
Increase in the frequency of heavy precipitation events in most areas	Very likely	Damage to crops; soil erosion; inability to cultivate land due to water logging of soils	Adverse effects on quality of surface and groundwater; contamination of water supply; water scarcity may be relieved	Increased risk of deaths; injuries; infectious, respiratory, and skin diseases	Disruption of settlements, commerce, transport, and societies due to flooding; pressures on urban and rural infrastructures; loss of property

(continues)

TABLE 6.2 (continued)
More IPCC Information on Projected Impacts and Their Likelihood

Phenomenon and Direction of Trend	Likelihood of Future Trends Based on Projections for 21st Century Using SRES Scenarios	Agriculture, Forestry, and Ecosystems	Water Resources	Human Health	Industry, Settlement, and Society
Increased frequency of precipitation events in most areas	Very likely	Damage to crops; soil erosion; inability to cultivate land due to water logging of soils	Adverse effects on quality of surface and groundwater; contamination of water supply; water scarcity may be relieved	Increased risk of deaths; injuries; infectious, respiratory, and skin diseases	Disruption of settlements, commerce, transport, and societies due to flooding, pressures on urban and rural infrastructures; loss of property
Area affected by drought increases	Likely	Land degradation, lower crop yields/crop damage and failure; increased livestock deaths; increased risk of wildfire	More widespread water stress	Increased risk of food and water shortage; increased risk of malnutrition; increased risk of water- and food-borne diseases	Water shortages for settlements, industry, and societies; reduced hydropower generation potentials; potential for population migrations
Intense tropical cyclone activity increases	Likely	Damage to crops; windthrow (uprooting) of trees; damage to coral reefs	Power outages cause disruption of public water supply	Increased risk of deaths; injuries; water- and food-borne diseases; posttraumatic stress disorders	Disruption by flood and high winds; withdrawal of risk coverage in vulnerable areas by private insurers; potential for population migrations; loss of property

TABLE 6.2 (continued)

More IPCC Information on Projected Impacts and Their Likelihood

Phenomenon and Direction of Trend	Likelihood of Future Trends Based on Projections for 21st Century Using SRES Scenarios	Agriculture, Forestry, and Ecosystems	Water Resources	Human Health	Industry, Settlement, and Society
Increased incidence of extreme high sea level (excludes tsunamis)	Likely	Salinization of irrigation water, estuaries, and freshwater systems	Decreased freshwater availability due to saltwater intrusion	Increased risk of deaths and injuries by drowning in floods; migration-related health effects	Costs of coastal protection versus cost of land-use relocation; potential for movement of population and infrastructure; also see tropical cyclones above

Note: SRES = **[Special Report on Emissions Scenarios]**.

IPCC 2000. "Special Report on Emissions Scenarios." Nebojsa Nakicenovic and Rob Swart [Eds.]. Cambridge University Press, UK. pp 570

Source: Adapted from Intergovernmental Panel on Climate Change. 2007. "Summary for Policymakers." In *Climate Change 2007: Impacts, Adaptation and Vulnerability. Contribution of Working Group II to the Fourth Assessment Report of the Intergovernmental Panel on Climate Change,* edited M. L. Parry, O. F. Canziani, J. P. Palutikof, P. J. van der Linden, and C. E. Hanson. Cambridge, UK: Cambridge University Press.

FIGURE 6.12

Estimated Deaths Attributed to Climate Change in the Year 2000 by Subregion

Mortality per Million
Population

0 – 2
2 – 4
4 – 70
70 – 120
no data

Note: Change in climate compared to baseline 1961–1990 climate.
Source: Reprinted by permission from Macmillan Publishers: Patz, Jonathan A., Diarmid Campbell-Lendrum, Tracey Holloway and Jonathan A. Foley. 2005. "Impact of Regional Climate Change on Human Health." *Nature* 438, 310–317.

FIGURE 6.13

Potential Economic Mitigation in 2030 by Region and Sector

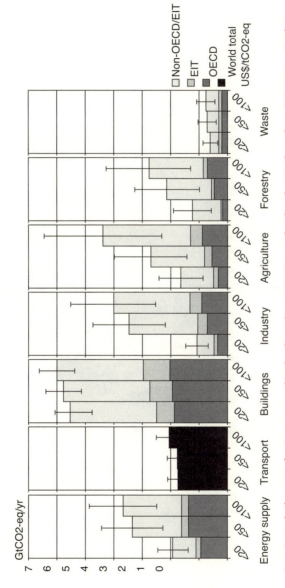

Source: Reprinted with permission from Intergovernmental Panel on Climate Change. 2007. "Summary for Policymakers." In *Climate Change 2007: Mitigation. Contribution of Working Group III to the Fourth Assessment Report of the Intergovernmental Panel on Climate Change,* edited by B. Metz, O. R. Davidson, P. R. Bosch, R. Dave, and L. A. Meyer. Cambridge, UK: Cambridge University Press.

FIGURES 6.14a & 6.14b
Estimated Economic Mitigation Potential in 2030

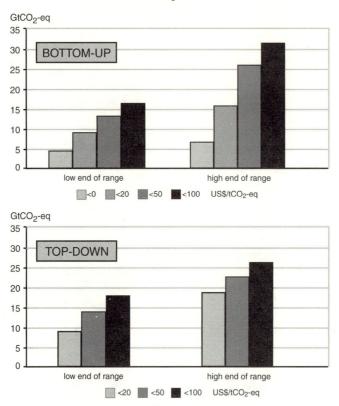

Source: Reprinted with permission from Intergovernmental Panel on Climate Control. 2007. "Summary for Policymakers." In *Climate Change 2007: Mitigation. Contribution of Working Group III to the Fourth Assessment Report of the Intergovernmental Panel on Climate Change,* edited by B. Metz, O. R. Davidson, P. R. Bosch, R. Dave, and L. A. Meyer. Cambridge, UK: Cambridge University Press.

are generally sectoral studies and take the macroeconomy as unchanged. Top-down approaches, on the other hand, assess the economy-wide potential of mitigation options using globally consistent frameworks and aggregated information.

Figure 6.15 shows the range of possible reductions to world gross domestic product (GDP) that may result from least-cost methods at different levels of climate change mitigation. The column entitled "Median GPD Reduction" lists expected reductions to global GDP (calculated using market exchange rates) by 2030; the rightmost column, entitled "Reduction of Average Annual GDP Growth Rates," is based on the average reduction during the period until 2030 that would result in the indicated GDP decrease in 2030.

Figure 6.16, taken from the *Stern Report on the Economics of Climate Change,* gives a picture of the possible costs of mitigation. According to Stern (2006), the average costs of carbon abatement are expected to decline by half over the next 20 years because of a number of factors, including technology development and transfer.

In order to stabilize emissions in the next 50 years, the world must reduce emissions by about 7 gigatons of carbon as compared to "business as usual" scenarios. Princeton University professors Rob Socolow and Stephen Pacala developed the "wedges" approach to frame the debate on how to achieve this ambitious goal. In a 2004 *Science* article, Pacala and Socolow identify 15 stabilization wedges that, if deployed at a significant global scale, could reduce emissions by 1 gigaton each. Wedges include actions such as increasing fuel economy from 30 to 60 miles per gallon for 2 billion cars, cutting carbon emission by 25 percent in buildings and appliances projected for 2054, and adding 4 million 1-megawatt-peak windmills. The approach is illustrated in Figure 6.17.

Figure 6.18 shows the geographical distribution of those industrialized countries with greenhouse gas emissions limitations under the Kyoto Protocol. These countries are referred to in the protocol as Annex B and are indicated on this map by their lighter color. The map helps to graphically show why discussions between the two groups are sometimes called north-south dialogues.

Table 6.3 lists adaptive capacities, vulnerabilities, and key concerns by region. It comes from the IPCC Third Assessment Report and underscores the ways in which climate change will impact different regions. It also highlights the wide range in the ability of regions to cope with climatic variability and change (list adapted from IPCC 2001a).

FIGURE 6.15
Macroeconomic Costs of Climate Change Mitigation in 2030

Trajectories toward Stabilization Levels (ppm CO_2-eq)	Median GDP Reduction (percentage points)	Range of GDP Reduction [1] (percentage points)	Reduction of Average Annual GDP Growth Rates (percentage points)
590–710	0.2	-0.6–1.2	< 0.06
535–590	0.6	0.2–2.5	< 0.1
445–535[2]	Not available	< 3	< 0.12

[1] The median and the 10th and 90th percentile range of the analyzed data are given.
[2] The number of studies that report GDP results is relatively small and they generally use low baselines.

Source: Intergovernmental Panel on Climate Change. 2007. "Summary for Policymakers." In *Climate Change 2007: Mitigation. Contribution of Working Group III to the Fourth Assessment Report of the Intergovernmental Panel on Climate Change,* edited by B. Metz, O. R. Davidson, P. R. Bosch, R. Dave, and L. A. Meyer. Cambridge, UK: Cambridge University Press.

FIGURE 6.16
Aggregate Cost Curve for Emissions Abatement

Cost of carbon abatement ($/tCO$_2$)

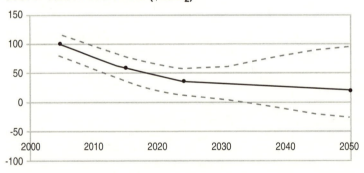

Source: Reprinted with permission from Stern, Nicholas. 2006. *Stern Report on the Economics of Climate Change.* Cambridge, UK: Cambridge University Press.

FIGURE 6.17
The Wedges

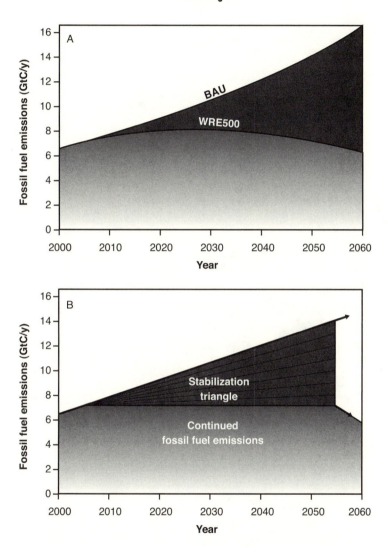

Source: Reprinted with permission from Pacala, S., and R. Socolow. 2004. "Stabilization Wedges: Solving the Climate Problem for the Next 50 Years with Current Technologies." [Online information; retrieved 7/22/08.] http://www.carbonsequestration.us/Papers-presentations/htm/Pacala-Socolow-ScienceMag-Aug2004.pdf.

FIGURE 6.18
Annex B and Non-Annex B Countries, 2005

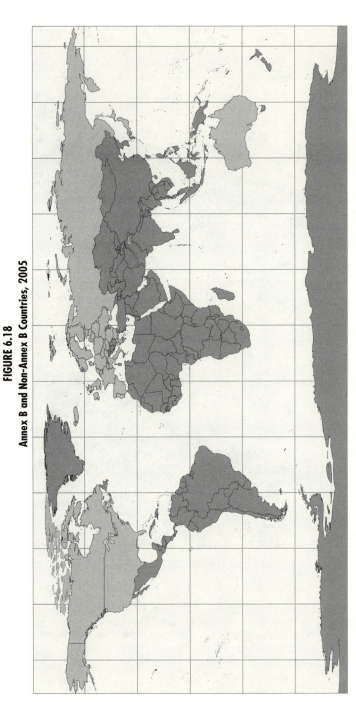

Source: Marland, G., and T. Boden. 2005. "Kyoto-Related Fossil-Fuel CO_2 Emission Totals." [Online information; retrieved 7/22/08.] http://cdiac.esd.ornl.gov/trends/emis/annex.htm

Table 6.4 presents examples of adaptation initiatives undertaken in response to present climate risks, including conditions associated with climate change (list adapted from Adger et al. 2007).

TABLE 6.3
Adaptive Capacities and Climate Vulnerabilities, by Region (adapted from IPCC 2001a)

Africa

- Adaptive capacity of human systems in Africa is low due to lack of economic resources and technology, and vulnerability is high as a result of heavy reliance on rain-fed agriculture, frequent droughts and floods, and poverty.
- Grain yields are projected to decrease for many scenarios, diminishing food security, particularly in small, food-importing countries (*medium to high confidence*).
- Major rivers of Africa are highly sensitive to climate variation; average runoff and water availability would decrease in the Mediterranean and countries of Southern Africa (*medium confidence*).
- Extension of ranges of infectious disease vectors would adversely affect human health in Africa (*medium confidence*).
- Desertification would be exacerbated by reductions in average annual rainfall, runoff, and soil moisture, especially in southern, North, and West Africa (*medium confidence*).
- Increases in droughts, floods, and other extreme events would add to stresses on water resources, food security, human health, and infrastructures and would constrain development in Africa (*high confidence*).
- Significant extinctions of plant and animal species are projected and would impact rural livelihoods, tourism, and genetic resources (*medium confidence*).
- Coastal settlements in, for example, the Gulf of Guinea, Senegal, Gambia, Egypt, and along the East-southern African coast would be adversely impacted by sea-level rise through inundation and coastal erosion (*high confidence*).

Asia

- Adaptive capacity of human systems is low and vulnerability is high in the developing countries of Asia; the developed countries of Asia are more able to adapt and less vulnerable.

(continues)

TABLE 6.3 (continued)
Adaptive Capacities and Climate Vulnerabilities, by Region (adapted from IPCC 2001a)

- Extreme events have increased in temperate and tropical Asia, including floods, droughts, forest fires, and tropical cyclones (*high confidence*).
- Decreases in agricultural productivity and aquaculture due to thermal and water stress, sea-level rise, floods and droughts, and tropical cyclones would diminish food security in many countries of arid, tropical, and temperate Asia; agriculture would expand and increase in productivity in northern areas (*medium confidence*).
- Runoff and water availability may decrease in arid and semi-arid Asia but increase in northern Asia (*medium confidence*).
- Human health would be threatened by possible increased exposure to vector-borne infectious diseases and heat stress in parts of Asia (*medium confidence*).
- Sea-level rise and an increase in the intensity of tropical cyclones would displace tens of millions of people in low-lying coastal areas of temperate and tropical Asia; increased intensity of rainfall would increase flood risks in temperate and tropical Asia (*high confidence*).
- Climate change would increase energy demand, decrease tourism attraction, and influence transportation in some regions of Asia (*medium confidence*).
- Climate change would exacerbate threats to biodiversity due to land-use and land-cover change and population pressure in Asia (*medium confidence*). Sea-level rise would put ecological security at risk, including mangroves and coral reefs (*high confidence*).
- Poleward movement of the southern boundary of the permafrost zones of Asia would result in a change of thermokarst and thermal erosion with negative impacts on social infrastructure and industries (*medium confidence*).

Australia and New Zealand

- Adaptive capacity of human systems is generally high, but groups in Australia and New Zealand, such as indigenous peoples in some regions, have low capacity to adapt and consequently high vulnerability.
- The net impact on some temperate crops of climate and CO_2 changes may initially be beneficial, but this balance is expected to become negative for some areas and crops with further climate change (*medium confidence*).

- Water is likely to be a key issue (*high confidence*) due to projected drying trends over much of the region and change to a more El Niño–like average state.
- Increases in the intensity of heavy rains and tropical cyclones (*medium confidence*) and region-specific changes in the frequency of tropical cyclones would alter the risks to life, property, and ecosystems from flooding, storm surges, and wind damage.
- Some species that occupy restricted climatic niches and that are unable to migrate due to fragmentation of the landscape, soil differences, or topography could become endangered or extinct (*high confidence*). Australian ecosystems that are particularly vulnerable to climate change include coral reefs, arid and semi-arid habitats in southwest and inland Australia, and Australian alpine systems. Freshwater wetlands in coastal zones in both Australia and New Zealand are vulnerable, and some New Zealand ecosystems are vulnerable to accelerated invasion by weeds.

Europe

- Adaptive capacity is generally high in Europe for human systems; southern Europe and the European Arctic are more vulnerable than other parts of Europe.
- Summer runoff, water availability, and soil moisture are likely to decrease in southern Europe and would widen the difference between the north and drought-prone south; increases are likely in winter in the north and south (*high confidence*).
- Half of alpine glaciers and large permafrost areas could disappear by end of the 21st century (*medium confidence*).
- River flood hazard will increase across much of Europe (*medium to high confidence*); in coastal areas, the risk of flooding, erosion, and wetland loss will increase substantially with implications for human settlement, industry, tourism, agriculture, and coastal natural habitats.
- Some broadly positive effects will be seen on agriculture in northern Europe (*medium confidence*); productivity will decrease in southern and eastern Europe (*medium confidence*).
- Upward and northward shift of biotic zones will take place. Loss of important habitats (wetlands, tundra, isolated habitats) would threaten some species (*high confidence*).
- Higher temperatures and heat waves may change traditional summer tourist destinations, and less reliable snow conditions may impact adversely on winter tourism (*medium confidence*).

(continues)

TABLE 6.3 (continued)
Adaptive Capacities and Climate Vulnerabilities, by Region (adapted from IPCC 2001a)

Latin America

- Adaptive capacity of human systems in Latin America is low, particularly with respect to extreme climate events, and vulnerability is high.
- Loss and retreat of glaciers would adversely impact runoff and water supply in areas where glacier melt is an important water source (*high confidence*).
- Floods and droughts would become more frequent, with floods increasing sediment loads and degrading water quality in some areas (*high confidence*).
- Increases in intensity of tropical cyclones would alter the risks to life, property, and ecosystems from heavy rain, flooding, storm surges, and wind damages (*high confidence*).
- Yields of important crops are projected to decrease in many locations in Latin America, even when the effects of CO_2 are taken into account; subsistence farming in some regions of Latin America could be threatened (*high confidence*).
- The geographical distribution of vector-borne infectious diseases would expand poleward and to higher elevations, and exposures to diseases such as malaria, dengue fever, and cholera will increase (*medium confidence*).

North America

- Adaptive capacity of human systems is generally high and vulnerability low in North America, but some communities (e.g., indigenous peoples and those dependent on climate-sensitive resources) are more vulnerable; social, economic, and demographic trends are changing vulnerabilities in subregions.
- Coastal human settlements, productive activities, infrastructure, and mangrove ecosystems would be negatively affected by sea-level rise (*medium confidence*).
- The rate of biodiversity loss would increase (*high confidence*).
- Some crops would benefit from modest warming accompanied by increasing CO_2, but effects would vary among crops and regions (*high confidence*), including declines due to drought in some areas of the Canadian Prairies and the U.S. Great Plains, potential increased food production in areas of Canada north of current production areas, and increased warm-temperate mixed forest production (*medium confidence*). However, benefits for crops would decline at an

increasing rate and possibly become a net loss with further warming (*medium confidence*).

- Snowmelt-dominated watersheds in western North America will experience earlier spring peak flows (*high confidence*), reductions in summer flows (*medium confidence*), and reduced lake levels and outflows for the Great Lakes–St. Lawrence region under most scenarios (*medium confidence*); adaptive responses would offset some, but not all, of the impacts on water users and on aquatic ecosystems (*medium confidence*).
- Unique natural ecosystems such as prairie wetlands, alpine tundra, and cold-water ecosystems will be at risk, and effective adaptation is unlikely (*medium confidence*).
- Sea-level rise would result in enhanced coastal erosion, coastal flooding, loss of coastal wetlands, and increased risk from storm surges, particularly in Florida and much of the U.S. Atlantic coast (*high confidence*).
- Weather-related insured losses and public-sector disaster relief payments in North America have been increasing; insurance-sector planning has not yet systematically included climate change information, so potential exists for surprise (*high confidence*).
- Vector-borne diseases—including malaria, dengue fever, and Lyme disease—may expand their ranges in North America; exacerbated air quality and heat stress morbidity and mortality would occur (*medium confidence*); socioeconomic factors and public health measures would play a large role in determining the incidence and extent of health effects.

Polar Regions

- Natural systems in polar regions are highly vulnerable to climate change, and current ecosystems have low adaptive capacity; technologically developed communities are likely to adapt readily to climate change, but some indigenous communities, in which traditional lifestyles are followed, have little capacity and few options for adaptation.
- Climate change in polar regions is expected to be among the largest and most rapid of any region on the Earth and will cause major physical, ecological, sociological, and economic impacts, especially in the Arctic, Antarctic Peninsula, and Southern Ocean (*high confidence*).
- Changes in climate that have already taken place are manifested in the decrease in extent and thickness of Arctic sea ice, permafrost thawing, coastal erosion, changes in ice

sheets and ice shelves, and altered distribution and abundance of species in polar regions (*high confidence*).
- Some polar ecosystems may adapt through eventual replacement by migration of species and changing species composition, and possibly by eventual increases in overall productivity; ice edge systems that provide habitat for some species would be threatened (*medium confidence*).
- Polar regions contain important drivers of climate change. Once triggered, they may continue for centuries, long after greenhouse gas concentrations are stabilized, and cause irreversible impacts on ice sheets, global ocean circulation, and sea-level rise (*medium confidence*).

Small Island States
- Adaptive capacity of human systems is generally low in small island states, and vulnerability is high; small island states are likely to be among the countries most seriously impacted by climate change.
- The projected sea-level rise of 5 millimeters per year[-1] for the next 100 years would cause enhanced coastal erosion, loss of land and property, dislocation of people, increased risk from storm surges, reduced resilience of coastal ecosystems, saltwater intrusion into freshwater resources, and high resource costs to respond and adapt to these changes (*high confidence*).
- Islands with very limited water supplies are highly vulnerable to the impacts of climate change on the water balance (*high confidence*).
- Coral reefs would be negatively affected by bleaching and by reduced calcification rates due to higher CO_2 levels (*medium confidence*); mangrove, sea grass bed, and other coastal ecosystems and the associated biodiversity would be adversely affected by rising temperatures and accelerated sea-level rise (*medium confidence*).
- Declines in coastal ecosystems would negatively impact reef fish and threaten reef fisheries, those who earn their livelihoods from reef fisheries, and those who rely on the fisheries as a significant food source (*medium confidence*).
- Limited arable land and soil salinization make agriculture of small island states, both for domestic food production and cash crop exports, highly vulnerable to climate change (*high confidence*).

- Tourism, an important source of income and foreign exchange for many islands, would face severe disruption from climate change and sea-level rise (*high confidence*).

TABLE 6.4
Adaptation Initiatives: Examples by Region (adapted from Adger et al. 2007)

Africa
Egypt
Sea-level rise: Adoption of National Climate Change Action Plan integrating climate change concerns into national policies; adoption of Law 4/94 requiring environmental impact assessment for project approval and regulating setback distances for coastal infrastructure; installation of hard structures in areas vulnerable to coastal erosion (El Raey 2004).

Sudan
Drought: Expanded use of traditional rainwater harvesting and water conserving techniques; building of shelter belts and windbreaks to improve resilience of rangelands; monitoring of the number of grazing animals and cut trees; setup of revolving credit funds (Osman-Elasha et al. 2006).

Botswana
Drought: National government programs to recreate employment options after drought; capacity building of local authorities; assistance to small subsistence farmers to increase crop production (FAO 2004).

Asia and Oceania
Bangladesh
Sea-level rise, saltwater intrusion: Consideration of climate change in the National Water Management Plan; building of flow regulators in coastal embankments; use of alternative crops and low-technology water filters (OECD 2003a; Pouliotte et al. 2006).

(continues)

Philippines

Drought, floods: Adjustment of silvicultural treatment schedules to suit climate variations; shift to drought-resistant crops; use of shallow tube wells; rotation method of irrigation during water shortage; construction of water-impounding basins; construction of fire lines and controlled burning; adoption of soil and water conservation measures for upland farming.

Sea-level rise, storm surges: Capacity building for shoreline defense system design; introduction of participatory risk assessment; provision of grants to strengthen coastal resilience and rehabilitation of infrastructures; construction of cyclone-resistant housing units; retrofit of buildings to improved hazard standards; review of building codes; reforestation of mangroves.

Drought, saltwater intrusion: Rainwater harvesting; leakage reduction; hydroponic farming; bank loans allowing for purchase of rainwater storage tanks (Lasco et al. 2006).

Americas
Canada

Permafrost melt; change in ice cover: Changes in livelihood practices by the Inuit, including change of hunt locations, diversification of hunted species, use of global positioning systems technology, encouragement of food sharing (Ford and Smit 2004).

Extreme temperatures: Implementation of heat health alert plans in Toronto, which include measures such as opening of designated cooling centers at public locations, information to the public through local media, distribution of bottled water through the Red Cross to vulnerable people, operation of a heat information line to answer heat-related questions, availability of an emergency medical service vehicle with specially trained staff and medical equipment (Mehdi 2006).

United States

Sea-level rise: Land acquisition programs taking account of climate change (e.g., New Jersey Coastal Blue Acres land acquisition programs to acquire coastal lands damaged/prone to damage by storms or buffering other lands (the acquired lands are being used for recreation and conservation); establishment of a "rolling easement" in Texas, an entitlement to public ownership of property that "rolls" inland with the coastline as the sea level rises; other coastal policies that encourage coastal landowners to act in ways that anticipate sea-level rise (Easterling, Hurd, and Smith 2004).

Mexico and Argentina
Drought: Adjustment of planting dates and crop variety (e.g., inclusion of drought-resistant plants such as agave and aloe); accumulation of commodity stocks as economic reserve; spatially separated lots for cropping and grazing to diversify exposures; diversification of income by adding livestock operations; setup/provision of crop insurance; creation of local financial pools (as alternative to commercial crop insurance) (Wehbe et al. 2006).

Europe
The Netherlands
Sea-level rise: Adoption of Flooding Defense Act and Coastal Defense Policy as precautionary approaches allowing for the incorporation of emerging trends in climate; building of a storm surge barrier taking a 50 centimeter sea-level rise into account; use of sand supplements added to coastal areas; improved management of water levels through dredging, widening of river banks, allowing rivers to expand into side channels and wetland areas; deployment of water storage and retention areas; conduct of regular (every five years) reviews of safety characteristics of all protecting infrastructure (dykes, etc.); preparation of risk assessments of flooding and coastal damage influencing spatial planning and engineering projects in the coastal zone; identifying areas for potential (land inward) reinforcement of dunes (Government of the Netherlands 1997, 2005).

Austria, France, Switzerland
Upward shift of natural snow reliability line, glacier melt: Artificial snowmaking; grooming of ski slopes; moving ski areas to higher altitudes and glaciers; use of white plastic sheets as protection against glacier melt; diversification of tourism revenues (e.g., all-year tourism).

(continues)

Permafrost melt, debris flows: Erection of protection dams in Pontresina, Switzerland, against avalanches and increased magnitude of potential debris flows stemming from permafrost thawing (Austrian Federal Government 2006; Direction du Tourisme 2002; Swiss Confederation 2005).

United Kingdom

Floods, sea-level rise: Coastal realignment under the Essex Wildlife Trust, converting more than 84 hectares of arable farmland into salt marsh and grassland to provide sustainable sea defenses; maintenance and operation of the Thames Barrier through the Thames Estuary 2100 project, which addresses flooding linked to the impacts of climate change; provision of guidance to policymakers, chief executives, and Parliament on climate change and the insurance sector (developed by the Association of British Insurers) (Defra 2006).

Documents

Excerpts from the Declaration of the United Nations (UN) Conference on the Human Environment (1972)

The 1972 UN Conference on the Human Environment was the first global conference specifically to address the environmental consequences of human activities. Among the many milestones the delegates achieved was a declaration recognizing that the natural environment provides the foundation of all human endeavors and calling on governments and individuals to take responsibility for protecting it. The declaration included many principles that have guided subsequent global efforts to protect the environment while preserving and encouraging human development, including the UN Framework Convention on Climate Change.

The declaration contains 7 proclamations and 26 principles. It states, in part:

The United Nations Conference on the Human Environment recognizes "the need for a common outlook and for common principles to inspire and guide the peoples of the world in the preservation and enhancement of the human environment."

It "calls upon Governments and peoples to exert common efforts for the preservation and improvement of the human environment, for the benefit of all the people and for their posterity."

It proclaims that "[T]o defend and improve the human environment for present and future generations has become an imperative goal for mankind—a goal to be pursued together with, and in harmony with, the established and fundamental goals of peace and of worldwide economic and social development."

It recognizes that "the developing countries must direct their efforts to development, bearing in mind their priorities and the need to safeguard and improve the environment. For the same purpose, the industrialized countries should make efforts to reduce the gap themselves and the developing countries."

Excerpts from the Final Report of the World Commission on Environment and Development (1983)

The World Commission on Environment and Development, commonly known as the Brundtland Commission, after Chair Gro Harlem Brundtland, was convened by the United Nations in 1983. Its modest remit was to develop an agenda for achieving sustainable development by 2000. In 1987, the commission released its report, Our Common Future, *urging the global community to embrace true multilateralism to address its most pressing challenges, all of which transcend national boundaries.*

The Brundtland report defines sustainable development as "development that meets the needs of the present without compromising the ability of future generations to meet their own needs."

It recognizes that "these challenges cut across the divides of national sovereignty, of limited strategies for economic gain, and of separated disciplines of science.

It calls for "decisive political action now to begin managing environmental resources to ensure both sustainable human progress and human survival. We are not forecasting a future; we are serving a notice—an urgent notice based on the latest and best scientific evidence—that the time has come to take the decisions needed to secure the resources to sustain this and coming generations. We do not offer a detailed blueprint for action, but instead a pathway by which the peoples of the world may enlarge their spheres of cooperation."

It states, "Failures to manage the environment and to sustain development threaten to overwhelm all countries. Environment and development are not separate challenges; they are inexorably linked. Development cannot subsist upon a deteriorating environmental resource base; the environment cannot be protected when growth leaves out of account the costs of environmental destruction. These problems cannot be treated separately by fragmented institutions and policies. They are linked in a complex system of cause and effect."

Excerpts from the Rio Declaration on Environment and Development (1992)

Produced at the 1992 United Nations Conference on Environment and Development, the Rio Declaration lays out 27 principles intended to guide future sustainable development around the world.

Principle 3 states, "The right to development must be fulfilled so as to equitably meet developmental and environmental needs of present and future generations."

Principle 4 states, "In order to achieve sustainable development, environmental protection shall constitute an integral part of the development process and cannot be considered in isolation from it."

Excerpts from the UN Framework Convention on Climate Change (UNFCCC)

Following release of the first report of the Intergovernmental Panel on Climate Change in 1990, the UN General Assembly officially authorized global negotiations intended to produce a framework treaty on climate change. The result was the UNFCCC, adopted in 1992, entering into force in 1994. To date, 191 countries, including the United States, and the European Community have ratified the UNFCCC.

Article 2 of the Convention describes its overall goals: "The ultimate objective of this Convention and any related legal instruments that the Conference of the Parties may adopt is to achieve, in accordance with the relevant provisions of the Convention, stabilization of greenhouse gas concentrations in the atmosphere at a level that would prevent dangerous anthropogenic interference with the climate system. Such a level should be achieved within a time frame sufficient to allow ecosystems to adapt naturally to climate change, to ensure that food production is not threatened and to enable economic development to proceed in a sustainable manner."

Article 3 also provides for differing obligations based on levels of development: "The parties should protect the climate system for the benefit of present and future generations of humankind, on the basis of equity and in accordance with their common but differentiated responsibilities and respective capabilities. Accordingly, the developed country Parties should take the lead in combating climate change and the adverse effects thereof."

Article 4 of the convention lays out obligations common to all parties:

"1. All Parties, taking into account their common but differentiated responsibilities and their specific national and regional development priorities, objectives and circumstances shall:

(a) Develop, periodically update, publish and make available . . . national inventories of anthropogenic emissions by sources and removal by sinks of all greenhouse gases not controlled by the Montreal Protocol. . . .

(b) Formulate, implement, publish and regularly update national, and where appropriate, regional programmes containing measures to mitigation climate change by addressing anthropogenic emissions by sources and sinks of all greenhouse gases not controlled by the Montreal Protocol, and measures to facilitate adequate adaptation to climate change;

2. The developed country Parties, and other Parties included in Annex I commit themselves specifically as provided for in the following:

(a) Each of these Parties shall adopt national policies and take corresponding measures on the mitigation of climate change, by limiting its anthropogenic emissions of greenhouse gases and protecting and enhancing its greenhouse gas sinks and reservoirs. These policies and measures will demonstrate that developed countries are taking the lead in modifying long-term trends in anthropogenic emissions consistent with the objective of the Convention. . . ."

Excerpts from the Kyoto Protocol

The Kyoto Protocol, a supplementary agreement to the UNFCCC, was adopted in 1997 and entered into force in 2005. Kyoto strengthens the convention by creating specific targets and timetables for the reduction of greenhouse gas emissions in Annex I (developed) countries. The protocol provides for action toward the goals of the convention, and the protocol is legally binding on its parties. To date, 181 parties have ratified the UNFCCC; the United States is not one of those countries.

Article 3 describes the commitments of the industrialized countries:

"1. The Parties included in Annex I shall, individually or jointly, ensure that their aggregate anthropogenic carbon dioxide equivalent emissions of the greenhouse gases . . . do not exceed their assigned amounts . . . with a view to reducing their overall emissions of such gases by at least 5 percent below 1990 levels in the commitment period 2008 to 2012.

2. Each Party included in Annex I shall, by 2005, have made demonstrable progress in achieving its commitments under this Protocol.

3. The net changes in greenhouse gas emissions by sources and removals by sinks resulting from direct human-induced land-use change and forestry activities, limited to afforestation, reforestation and deforestation since 1990,

measured as verifiable changes in carbon stocks in each commitment period, shall be used to meet the commitments under this Article of each Party included in Annex I. The greenhouse gas emissions by sources and removals by sinks associated with those activities shall be reported in a transparent and verifiable manner and reviewed in accordance with Articles 7 and 8."

Article 10 describes commitments for all parties:
"All Parties, taking into account their common but differentiated responsibilities and their specific national and regional development priorities, objectives and circumstances . . . shall:

(a) Formulate . . . cost-effective . . . programmes to improve the quality of local emission factors, activity data and/or models which reflect the socio-economic conditions of each Party for the preparation and periodic updating of national inventories of anthropogenic emissions by sources and removals by sinks of all greenhouse gases not controlled by the Montreal Protocol . . . ;

(b) Formulate, implement, publish and regularly update national and, where appropriate, regional programmes containing measures to mitigate climate change and measures to facilitate adequate adaptation to climate change:

(i) Such programmes would, inter alia, concern the energy, transport and industry sectors as well as agriculture, forestry and waste management. Furthermore, adaptation technologies and methods for improving spatial planning would improve adaptation to climate change; and

(ii) Parties included in Annex I shall submit information on action under this Protocol, including national programmes, in accordance with Article 7; and other Parties shall seek to include in their national communications, as appropriate, information on programmes which contain measures that the Party believes contribute to addressing climate change and its adverse impacts, including the abatement of increases in greenhouse gas emissions, and enhancement of and removals by sinks, capacity building and adaptation measures;

(c) Cooperate in the promotion of effective modalities for the development, application and diffusion of, and take all practicable steps to promote, facilitate and finance, as appropriate, the transfer of, or access to, environmentally sound technologies, know-how, practices and processes pertinent to climate change, in particular to developing countries, including the formulation of policies and programmes for the effective transfer of environmentally sound technologies that are publicly owned or in the public domain and the creation of an enabling environment for the private sector, to promote and enhance the transfer of, and access to, environmentally sound technologies;

(d) Cooperate in scientific and technical research and promote the maintenance and the development of systematic observation systems and development of data archives to reduce uncertainties related to the climate system, the adverse impacts of climate change and the economic and social consequences of various response strategies, and promote the development and strengthening of endogenous capacities and capabilities to participate in international and intergovernmental efforts, programmes and networks on research and systematic observation, taking into account Article 5 of the Convention;

(e) Cooperate in and promote at the international level, and, where appropriate, using existing bodies, the development and implementation of education and training programmes, including the strengthening of national capacity building, in particular human and institutional capacities and the exchange or secondment of personnel to train experts in this field, in particular for developing countries, and facilitate at the national level public awareness of, and public access to information on, climate change. Suitable modalities should be developed to implement these activities through the relevant bodies of the Convention, taking into account Article 6 of the Convention;

(f) Include in their national communications information on programmes and activities undertaken pursuant to this Article in accordance with relevant decisions of the Conference of the Parties . . .

Excerpts from IPCC Reports

Working Group I: The Science of Climate Change. Excerpts from the Assessment Reports of Working Group I (the Scientific Basis) of the Intergovernmental Panel on Climate Change (IPCC WG 1)

WG I—Second Assessment Report, issued 1995:
"The balance of evidence suggests a discernible human influence on global climate" (IPCC 1995, 2).

WG I—Third Assessment Report, issued 2001:
"There is new and stronger evidence that most of the warming of the last 50 years is attributable to human activities" (IPCC 2001b, 9).

WG I—Fourth Assessment Report, issued 2007:
"Warming of the climate system is unequivocal" (IPCC 2007a, 5).

"Most of the observed increase in globally averaged temperatures since the mid-20th century is very likely due to the observed increase in anthropogenic greenhouse gas concentrations" (IPCC 2007a, 10).

Working Group II: Impacts, Adaptation and Vulnerability. Excerpts from the Assessment Reports of Working Group II (Impacts, Adaptation and Vulnerability) of the Intergovernmental Panel on Climate Change (IPCC WG II).

WG II—Second Assessment Report, issued 1995:
"The reliability of regional scale predictions is still low and the degree to which climate variability may change is uncertain. However, potentially serious changes have been identified, including an increase in some regions in the incidence of extreme high temperature events, floods and droughts, with resultant consequences for fires, pest outbreaks, and ecosystem composition, structure and functioning, including primary productivity" (Watson, Zinyowera, and Moss 1995, 2).

WG II—Third Assessment Report, issued 2001:
"From the collected evidence, there is *high confidence* that recent regional changes in temperature have had discernible impacts on many physical and biological systems" (IPCC 2001c, 4).

"It is *well-established* that the geographic extent of the damage or loss [due to climate change] and the number of systems affected will increase with the magnitude and rate of climate change" (IPCC 2001c, 5).

WG II—Fourth Assessment Report, issued 2007:

"Observational evidence from all continents and most oceans shows that many natural systems are being affected by regional climate changes, particularly temperature increases" (IPCC 2007b, 8).

"Other effects of regional climate changes on natural and human environments are emerging, although many are difficult to discern due to adaptation and non-climatic drivers" (IPCC 2007b, 9).

"Some large-scale climate events have the potential to cause very large impacts, especially after the 21st century" (IPCC 2007b, 17), and "adaptation will be necessary to address impacts resulting from the warming which is already unavoidable due to past emissions" (IPCC 2007b, 19).

Working Group III: Mitigation of Climate Change. Excerpts from the Assessment Reports of Working Group III (Mitigation) of the Intergovernmental Panel on Climate Change (IPCC WG III)

WG III—Fourth Assessment Report, issued 2007:

"In order to stabilize the concentration of GHGs [greenhouse gases] in the atmosphere, emissions would need to peak and decline thereafter. The lower the stabilization level, the more quickly this peak and decline would need to occur. Mitigation efforts over the next two to three decades will have a large impact on opportunities to achieve lower stabilization levels" (IPCC 2007c, 15).

"The range of stabilization levels assessed can be achieved by deployment of a portfolio of technologies that are currently available and those that are expected to be commercialised in coming decades. This assumes that appropriate and effective incentives are in place for development, acquisition, deployment and diffusion of technologies and for addressing related barriers" (IPCC 2007c, 16).

"Both bottom-up and top-down studies indicate that there is substantial economic potential for the mitigation of global GHG emissions over the coming decades, that could offset the projected growth of global emissions or reduce emissions below current levels" (IPCC 2007c, 9).

"New energy infrastructure investments in developing countries, upgrades of energy infrastructure in industrialized countries, and policies that promote energy security, can, in many cases, create opportunities to achieve GHG emission reductions compared to baseline scenarios. Additional co-benefits are country-specific but often include air pollution abatement, balance of trade improvement, [and] provision of modern energy services to rural areas and employment" (IPCC 2007c, 12).

Byrd-Hagel Resolution (1997)

S. Res. 98 Expressing the sense of the Senate regarding the conditions for the United States becoming a signatory to any international agreement on greenhouse gas emissions under the United Nations (U.S. Congress 1997).

"Whereas the exemption for Developing Country Parties is inconsistent with the need for global action on climate change and is environmentally flawed;

Whereas the Senate strongly believes that the proposals under negotiation, because of the disparity of treatment between Annex I Parties and Developing Countries and the level of required emission reductions, could result in serious harm to the United States economy, including significant job loss, trade disadvantages, increased energy and consumer costs, or any combination thereof; and

Whereas it is desirable that a bipartisan group of Senators be appointed by the Majority and Minority Leaders of the Senate for the purpose of monitoring the status of negotiations on Global Climate Change and reporting periodically to the Senate on those negotiations: Now, therefore, be it

Resolved, That it is the sense of the Senate that—

(1) the United States should not be a signatory to any protocol to, or other agreement regarding, the United Nations Framework Convention on Climate Change of 1992, at negotiations in Kyoto in December 1997, or thereafter, which would—

(a) mandate new commitments to limit or reduce greenhouse gas emissions for the Annex I Parties, unless the protocol or other agreement also mandates new specific scheduled commitments to limit or reduce greenhouse gas emissions for Developing Country Parties within the same compliance period, or

(b) would result in serious harm to the economy of the United States; and

(2) any such protocol or other agreement which would require the advice and consent of the Senate to ratification should be accompanied by a detailed explanation of any legislation or regulatory actions that may be required to implement the protocol or other agreement and should also be accompanied by an analysis of the detailed financial costs and other impacts on the economy of the United States which would be incurred by the implementation of the protocol or other agreement.

Joint Statement of the National Science Academies of the G8 (G8 2005)

The national science academies of the Group of Eight (G8) (Canada, France, Germany, Italy, Japan, Russia, the United Kingdom, and the United States) nations and Brazil, China, and India, three of the largest emitters of greenhouse gases in the developing world, issued a statement on June 7, 2005, calling on world leaders, particularly those of the G8 countries, to acknowledge that the threat of climate change is clear and increasing, to address its causes, and to prepare for its consequences.

It states, "The scientific understanding of climate change is now sufficiently clear to justify nations taking prompt action. It is vital that all nations identify cost-effective steps that they can take now, to contribute to substantial and long-term reduction in net global greenhouse gas emissions."

And "We urge all nations, in the line with the UNFCCC principles, to take prompt action to reduce the causes of climate change, adapt to its impacts and ensure that the issue is included in all relevant national and international strategies. As national science academies, we commit to working with governments to help develop and implement the national and international response to the challenge of climate change.

G8 nations have been responsible for much of the past greenhouse gas emissions. As parties to the UNFCCC, G8 nations are committed to showing leadership in addressing climate change and assisting developing nations to meet the challenges of adaptation and mitigation.

We call on world leaders, including those meeting at the Gleneagles G8 Summit in July 2005, to:

> Acknowledge that the threat of climate change is clear and increasing.

> Launch an international study to explore scientifically informed targets for atmospheric greenhouse gas concentrations, and their associated emissions scenarios, that will enable nations to avoid impacts deemed unacceptable.

> Identify cost-effective steps that can be taken now to contribute to substantial and long-term reduction in net global greenhouse gas emissions. Recognise that delayed action will increase the risk of adverse environmental effects and will likely incur a greater cost.

> Work with developing nations to build a scientific and technological capacity best suited to their circumstances, enabling them to develop innovative solutions to mitigate and adapt to the adverse effects of climate change, while explicitly recognising their legitimate development rights.

> Show leadership in developing and deploying clean energy technologies and approaches to energy efficiency, and share this knowledge with all other nations.

> Mobilise the science and technology community to enhance research and development efforts, which can better inform climate change decisions."

Climate Change: An Evangelical Call to Action

Released by a coalition of leaders of American evangelical Christian churches, this statement presents action on climate change as a moral imperative for Christians.

It reads in part:

"Over the last several years many of us have engaged in study, reflection, and prayer related to the issue of climate change (often called "global warming"). For most of us, until recently this has not been treated as a pressing issue or major priority. Indeed, many of us have required considerable convincing before becoming persuaded that climate change is a real problem and that it ought to matter to us as Christians" (Evangelical Climate Initiative 2006, 2).

"In the face of the breadth and depth of this scientific and governmental concern [with regard to climate change] . . . we are convinced that evangelicals must engage this issue without any further lingering over the basic reality of the problem or humanity's responsibility to address it" (Evangelical Climate Initiative 2006, 3).

"Love of God, love of neighbor, and the demands of stewardship are more than enough reason for evangelical Christians to respond to the climate change problem with moral passion and concrete action" (Evangelical Climate Initiative 2006, 6).

References

Adger, W. N., S. Agrawala, M. M. Q. Mirza, C. Conde, K. O'Brien, J. Pulhin, R. Pulwarty, B. Smit, and K. Takahashi, 2007. "Assessment of Adaptation Practices, Options, Constraints and Capacity." In *Climate Change 2007: Impacts, Adaptation and Vulnerability. Contribution of Working Group II to the Fourth Assessment Report of the Intergovernmental Panel on Climate Change*, edited by M. L. Parry, O. F. Canziani, J. P. Palutikof, P. J. van der Linden, and C. E. Hanson, 717–743. Cambridge, UK: Cambridge University Press.

Austrian Federal Government. 2006. *Fourth National Communication of the Austrian Federal Government, in Compliance with the Obligations under the United Nations Framework Convention on Climate Change (Federal Law Gazette No. 414/1994), according to Decisions 11/CP.4 and 4/CP.5 of the Conference of the Parties.* Vienna: Federal Ministry of Agriculture, Forestry, Environment and Water Management, Vienna.

Brundtland, G. H., Khalid, M., Agnelli, S., Al-Athel, S.A., Chidzero, B., Fadika, L.M., Hauff, V., Lang, I., Shijun, M. Marino de Botero, M., Singh, N., Nogueira-Neto, P., Okita, S., Ramphal, S.S., Ruckelshaus, W.D., Sahnoun, M., Salim, E., Shaib, B., Sokolov, V., Stanounik, J., and Strong, M. (1987) *Report of the World Commission on Environment and Development:*

Our Common Future. [Online report; retrieved 9/17/07.] http://www
.un-documents.net/wced-ocf.htm.

Direction du Tourisme. 2002. *"Les chiffres clés du tourisme de montagne en
France, 3ème edition."* Paris: Service d'Etudes et d'Aménagement touris-
tique de la montagne.

Easterling, W. E., B. H. Hurd, and J. B. Smith. 2004: "Coping with Global
Climate Change: The Role of Adaptation in the United States." Arlington, VA: Pew Center on Global Climate Change.

El Raey, M. 2004. *Adaptation to Climate Change for Sustainable Development
in the Coastal Zone of Egypt.* Report No. ENV/EPOC/GF/SD/
RD(2004)1/FINAL. Paris: Organisation for Economic Co-operation and
Development.

Evangelical Climate Initiative. 2006. "Climate Change: An Evangelical
Call to Action." [Online statement; retrieved 5/22/08.]
http://christiansandclimate.org/learn/call-to-action/.

Food and Agriculture Organization (FAO), 2004: *Drought Impact Mitigation and Prevention in the Limpopo River Basin: A Situation Analysis.* Land
and Water Discussion Paper 4. Rome: FAO.

Ford, J., and B. Smit. 2004. "A Framework for Assessing the Vulnerability
of Communities in the Canadian Arctic to Risks Associated with Climate
Change." *Arctic* 57:389–400.

Government of the Netherlands. 1997. *Second Netherlands' Communication
on Climate Change Policies.* Prepared for the Conference of Parties under
the Framework Convention on Climate Change. The Hague: Ministry of
Housing, Spatial Planning and the Environment, Ministry of Economic
Affairs, Ministry of Transport, Public Works and Water Management,
Ministry of Agriculture, Nature Management and Fisheries, Ministry of
Foreign Affairs.

Government of the Netherlands. 2005. *Fourth Netherlands' National Communication under the United Nations Framework Convention on Climate
Change.* The Hague: Ministry of Housing, Spatial Planning and the Environment.

Group of Eight (G8) Science Academies. 2005. "Joint Science Academies'
Statement: Global Response to Climate Change." [Online information;
retrieved 7/14/08.] http://www.scj.go.jp/ja/info/kohyo/pdf/kohyo-
19-s1027.pdf.

Intergovernmental Panel on Climate Change. 1995. "Summary for Policymakers." In *Contribution of Working Group I to the Second Assessment of
the Intergovernmental Panel on Climate Change,* edited by J. T. Houghton,

L. G. Meira Filho, B. A. Callender, N. Harris, A. Kattenberg, and K. Maskell. Cambridge, UK: Cambridge University Press.

Intergovernmental Panel on Climate Change (IPCC). 2001a. "Summary for Policymakers." In *Climate Change 2001: Impacts, Adaptation and Vulnerability. Contribution of Working Group II to the Third Assessment Report of the Intergovernmental Panel on Climate Change,* edited by J. J. McCarthy, O. F. Canziani, N. A. Leary, D. J. Dokken, and K. S. White. Cambridge, UK: Cambridge University Press.

Intergovernmental Panel on Climate Change (IPCC). 2001b. "Summary for Policymakers." In *Climate Change 2001: The Scientific Basis. Contribution of Working Group I to the Third Assessment Report of the Intergovernmental Panel on Climate Change,* edited by J. T. Houghton, Y. Ding, D. J. Griggs, M. Noguer, P. J. van der Linden, X. Dai, K. Maskell, and C. A. Johnson. Cambridge, UK: Cambridge University Press.

Intergovernmental Panel on Climate Change (IPCC). 2001c. "Summary for Policymakers." In *Climate Change 2001: Impacts, Adaptation and Vulnerability. Contribution of Working Group II to the Third Assessment Report of the Intergovernmental Panel on Climate Change,* edited by J. J. McCarthy, O. F. Canziani, N. A. Leary, D. J. Dokken, and K. S. White. Cambridge, UK: Cambridge University Press.

Intergovernmental Panel on Climate Change (IPCC). 2007a. "Summary for Policymakers." In *Climate Change 2007: The Physical Science Basis. Contribution of Working Group I to the Fourth Assessment Report of the Intergovernmental Panel on Climate Change,* edited by S. Solomon, D. Qin, M. Manning, Z. Chen, M. Marquis, K. B. Averyt, M. Tignor, and H. L. Miller. Cambridge, UK: Cambridge University Press.

Intergovernmental Panel on Climate Change (IPCC). 2007b. "Summary for Policymakers." In *Climate Change 2007: Impacts, Adaptation and Vulnerability. Contribution of Working Group II to the Fourth Assessment Report of the Intergovernmental Panel on Climate Change,* edited by M. L. Parry, O. F. Canziani, J. P. Palutikof, P. J. van der Linden, and C. E. Hanson, 7–22. Cambridge, UK: Cambridge University Press.

Intergovernmental Panel on Climate Change (IPCC). 2007c. "Summary for Policymakers." In *Climate Change 2007: Mitigation. Contribution of Working Group III to the Fourth Assessment Report of the Intergovernmental Panel on Climate Change,* edited by B. Metz, O. R. Davidson, P. R. Bosch, R. Dave, and L. A. Meyer. Cambridge, UK: Cambridge University Press.

Lasco, R., R. Cruz, J. Pulhin, and F. Pulhin. 2006. "Tradeoff Analysis of Adaptation Strategies for Natural Resources, Water Resources and Local Institutions in the Philippines." Assessments of Impacts and Adaptations to Climate Change Working Paper No. 32. Washington, DC: International START Secretariat.

Mehdi, B. 2006. *Adapting to Climate Change: An Introduction for Canadian Municipalities.* Vancouver BC: Canadian Climate Impacts and Adaptation Research Network (C-CIARN), 32 p.

Organisation for Economic Co-operation and Development (OECD). 2003. "Development and Climate Change in Bangladesh: Focus on Coastal Flooding and the Sundarbans." Report No. COM/ENV/EPOC/ DCD/DAC(2003) 3/FINAL. Paris: OECD.

Osman-Elasha, B., N. Goutbi, E. Spanger-Siegfried, B. Dougherty, A. Hanafi, S. Zakieldeen, A. Sanjak, H. Atti, and H. Elhassan. 2006. "Adaptation Strategies to Increase Human Resilience against Climate Variability and Change: Lessons from the Arid Regions of Sudan." Assessments of Impacts and Adaptations to Climate Change Working Paper 42. Washington, DC: International START Secretariat.

Pacala, S., and R. Socolow. 2004. "Stabilization Wedges: Solving the Climate Problem for the Next 50 Years with Current Technologies." [Online article; retrieved 7/22/08.] http://www.carbonsequestration.us/ Papers-presentations/htm/Pacala-Socolow-ScienceMag-Aug2004.pdf.

Pouliotte, J., N. Islam, B. Smit, and S. Islam. 2006. "Livelihoods in Rural Bangladesh." [Online article; retrieved 10/11/06.] http://www.cru.uea .ac.uk/tiempo/portal/archive/pdf/tiempo59low.pdf.

Stern, Nicholas. 2006. *Stern Report on the Economics of Climate Change.* Cambridge, UK: Cambridge University Press.

Swiss Confederation. 2005. *Switzerland's Fourth National Communication under the UNFCCC.* Berne: Swiss Agency for the Environment, Forests and Landscape.

UK Department for Environment, Food and Rural Affairs. 2006. *The UK's Fourth National Communication under the United Nations Framework Convention on Climate Change.* London: UK Department for Environment, Food and Rural Affairs.

United Nations Conference on Environment and Development. 1992. "Rio Declaration on Environment and Development." [Online report; retrieved 8/8/07.] http://www.un.org/documents/ga/conf151/aconf15 126-1annex1.htm.

United Nations Environment Programme. 1972. "Declaration of the United Nations Conference on the Human Environment." [Online information; retrieved 7/22/08.] http://www.unep.org/Documents .Multilingual/Default.asp?DocumentID=97&ArticleID=1503.

United Nations Framework Convention on Climate Change. 1992. *United Nations Framework Convention on Climate Change.* [Online information; retrieved 7/21/07.] http://unfccc.int/not_assigned/b/items/1417.php.

United Nations Framework Convention on Climate Change. 1997. *Kyoto Protocol to the United Nations Framework Convention on Climate Change.* [Online information; retrieved 7/17/07.] http://unfccc.int/essential_ background/kyoto_protocol/items/1678.php.

U.S. Congress. 1997. "Byrd-Hagel Resolution." [Online U.S. Senate resolution; retrieved 7/14/08.] www.nationalcenter.org/KyotoSenate.html.

Watson, R. T., M. C. Zinyowera, and R. H. Moss. 1995. "Summary for Policymakers." In *Climate Change 1995: Impacts, Adaptations and Mitigation of Climate Change: Scientific-Technical Analyses,* edited by R. T. Watson, M. C. Zinyowera and R. H. Moss. Cambridge, UK: Cambridge University Press.

Wehbe, M., H. Eakin, R. Seiler, M. Vinocur, C. Afila, and C. Marutto. 2006. "Local Perspectives on Adaptation to Climate Change: Lessons from Mexico and Argentina." Assessments of Impacts and Adaptations to Climate Change Working Paper 39. Washington, DC: International START Secretariat.

7

Directory of Organizations

This chapter provides basic information about some of the most important organizations involved in climate science and policy. The chapter divides the organizations into several groups: treaty secretariats, international organizations (IOs), governmental agencies, research centers, environmental nongovernmental organizations (ENGOs), and business organizations.

Climate change is a huge issue and one that intersects with nearly all other issue areas. Consequently, space constraints do not allow us to list every intergovernmental organization (IGO), national agency, nongovernmental organization (NGO), business group, or research center that works in this area. However, we have tried to include the most prominent or important institutions as well as other organizations that possess particular types of expertise, play important roles on certain related issues, or maintain particularly useful Web sites. At the same time, we recognize that other organizations would merit listing if space allowed.

We also recognize that some organizations fit into more than one category. For example, the World Conservation Union (IUCN), which we list as an IGO, includes both national governments and NGOs as members. Far more overlap exists among the different types of nongovernmental organizations. Many of the environmental and business-oriented NGOs conduct and publish research; some research organizations are so closely associated with environmental goals that they might be considered environmental organizations (e.g., the World Resources Institute); and some of the business-oriented NGOs have taken positions in favor of strong environmental policy. Thus, while we have tried to place each organization in the most appropriate category, we recognize that some could be classified differently.

237

This directory seeks to help guide you as you search for more detailed information. Each listing includes a very brief description of the group's mission as it relates to climate change as well as contact information and Web addresses. The Web sites listed here provide links to all of the primary source documents referred to in this book (such as the climate treaties and Intergovernmental Panel on Climate Change, or IPCC, reports), as well as a tremendous array of other documents, reports, newsletters, press releases, news stories, databases, and other materials. We include e-mail addresses for organizations that maintain general addresses for information inquiries. For those that do not, please consult their Web site to find the appropriate department for your needs. Please also note that we have included telephone country codes for organizations headquartered outside the United States and Canada but that the international dialing prefix 011 must be used when calling most overseas numbers from the United States.

Treaty Secretariats

Most global environmental treaties contain provisions establishing a secretariat to manage day-to-day treaty operations; organize meetings; gather information; create reports; act as liaisons to other treaty secretariats and organizations; provide information to parties (countries that have officially ratified and joined the treaty), the media, and the public; and perform other functions. This section lists secretariats for several of the most prominent global environmental treaties. Each entry includes a brief description of the treaty and contact information for the secretariat.

Those seeking information on other environmental treaties, and there are hundreds, can consult the complete list of links to all environmental treaty secretariats, divided by issue area, maintained by the International Institute for Sustainable Development at http://www.iisd.ca/process/weblinks.htm. The text of most environmental treaties can be found on an excellent, searchable, online database maintained by the Center for International Earth Science Information Network at http://sedac.ciesin.org/entri/.

Convention on Biological Diversity (CBD)
Secretariat of the Convention on Biological Diversity
413 Saint Jacques Street, Suite 800
Montreal QC H2Y 1N9, Canada
E-mail: secretariat@cbd.int
Web site: http://www.cbd.int

The goal of the 1992 Convention on Biological Diversity is to help governments develop national policies for the conservation and sustainable use of biological diversity. It stands at the center of a cluster of treaties that seek to protect endangered species and their habitats. The Convention also seeks to ensure fair and equitable sharing of the benefits of genetic resources. The CBD entered into force on December 29, 1993. As of August 2008, there were 191 parties. The United States is not among them.

As climate change could lead to a significant increase in extinctions, biodiversity loss is closely linked to climate change. Consequently, the intersection of climate change and biodiversity is a cross-cutting theme under the CBD. The Web site contains information on activities being undertaken to implement this theme and integrate climate change concerns into CBD programs. One example of this effort is the Adaptation Database, which provides case studies of the integration of biodiversity concerns into planning for adaptation to climate change. Other technical papers hosted on the site provide lessons learned from these case studies.

Convention on Long-Range Transboundary Air Pollution (CLRTAP)
United Nations Economic Commission for Europe
Environment and Human Settlements Division
Palais des Nations
CH-1211 Geneva, Switzerland
E-mail: air.env@unece.org
Web site: http://www.unece.org/env/lrtap/

The Convention on Long-Range Transboundary Air Pollution is a regional treaty covering Europe, Russia, and North America that seeks to protect human health and the environment from air pollution. The treaty calls for the gradual reduction and prevention of air pollution, including long-range transboundary air pollution, through the sharing of information and strategies between parties and the creation of protocols that enact binding controls for specific pollutants. CLRTAP entered into force on March 16, 1983. To date, eight protocols have been added to the convention, including those that cover ground-level ozone (1999), persistent organic pollutants (1998), heavy metals (1998), sulfur emissions (1994), volatile organic compounds (1991), and nitrogen oxides (1988). A list of parties to each protocol is available on the treaty's Web site.

Kyoto Protocol to the United Nations Framework Convention on Climate Change (UNFCCC)
UNFCCC Secretariat
P.O. Box 260124
D-53153 Bonn, Germany
E-mail: secretariat@unfccc.int
Web site: http://unfccc.int/kyoto_protocol/items/2830.php

The 1997 Kyoto Protocol is the instrument added to the 1992 United Nations Framework Convention on Climate Change that mandates further action to address climate change. The protocol shares the UNFCCC's objective, principles, and institutions but significantly strengthens the Convention by committing Annex I parties (i.e., industrialized countries) to individual, legally binding targets to limit or reduce their greenhouse gas emissions and creating mechanisms to allow parties flexibility in how they achieve these reductions. The targets, which are different for each country, add up to a total cut in greenhouse gas emissions of at least 5 percent from 1990 levels during the period 2008–2012. Only parties to the Convention that have also become parties to the Protocol (i.e., by ratifying, accepting, approving, or acceding to it) are bound by the Protocol's commitments. The Protocol entered into force on February 16, 2005. As of August 2008, there were 182 parties. The United States is not a party.

Montreal Protocol on Substances That Deplete the Ozone Layer
Ozone Secretariat
P.O. Box 30552
Nairobi, Kenya
E-mail: ozoneinfo@unep.org
Web site: http://ozone.unep.org

The 1987 Montreal Protocol stands at the heart of the ozone regime—the series of global agreements and international institutions designed to protect stratospheric ozone. Stratospheric ozone, commonly known as the ozone layer, helps shield the

Earth from harmful ultraviolet radiation from the sun. The Protocol is a specific off-shoot of the 1985 Vienna Convention for the Protection of the Ozone Layer, which is a framework convention similar to the UNFCCC. The Montreal Protocol, perhaps the most influential and successful global environmental treaty in history, has been strengthened several times through amendments and now requires all parties to eliminate or severely restrict production and use of all chemicals known to damage the ozone layer, including chlorofluorocarbons (CFCs). The Montreal Protocol entered into force on January 1, 1989. As of August 2008, there were 193 parties to the Protocol. The Ozone Secretariat administers both the Vienna Convention and the Montreal Protocol and is housed at the United Nations Environment Programme's (UNEP) headquarters in Nairobi. A separate secretariat administers the Multilateral Fund for the Implementation of the Montreal Protocol (http://www.multilateralfund.org/), which provides resources to developing countries to help them eliminate the production and use of CFCs and other ozone-depleting chemicals.

The Web site contains a great deal of information, including issues related to the intersection of ozone and climate concerns. Such information includes the recent agreement by parties to accelerate the phase-out of hydrochlorofluorocarbons, which are one of the replacement chemicals for CFCs but which are also potent greenhouse gases.

Stockholm Convention on Persistent Organic Pollutants
Secretariat for the Stockholm Convention
United Nations Environment Programme
International Environment House
11–13 chemin des Anemones
CH-1219 Chatelaine, Geneva, Switzerland
E-mail: ssc@pops.int
Web site: http://www.pops.int

The 2001 Stockholm Convention seeks to protect human health and the environment from persistent organic pollutants (POPs). POPs are a suite of substances found to possess four key characteristics that make them extremely dangerous: They are highly toxic; they can travel long distances from their point of emission via air and water currents, via patterns of evaporation and rainfall, and within the bodies of migratory species; they can stay in the environment for long periods of time before they break down (i.e., they possess a high level of persistence due to long half-lives); and they can bioaccumulate in the tissues of living organisms, concentrating their toxic impact over time even if individual exposures are quite small. The Stockholm Convention requires that countries eliminate or severely restrict the production, use, and release of 12 specific POPs and includes a provision for adding to this list as needed. The Stockholm Convention entered into force on May 17, 2004. As of August 2008, there were 156 parties. The United States is not a party.

United Nations Convention to Combat Desertification (UNCCD)
UNCCD Secretariat
P.O. Box 260129
D-53153 Bonn, Germany
E-mail: secretariat@unccd.int
Web site: http://www.unccd.int

The United Nations Convention to Combat Desertification seeks to fight desertification and mitigate the effects of drought, particularly in Africa, through long-term strategies that improve land productivity and encourage sustainable management of land and water resources. The Convention entered into force on December 26, 1996. As of August 2008, there were 193 parties to the Convention. As climate change is expected to exacerbate desertification in some regions, this regime has received increased attention from those interested in assisting arid countries to prepare for climate change.

United Nations Framework Convention on Climate Change (UNFCCC)
UNFCCC Secretariat
P.O. Box 260124
D-53153 Bonn, Germany
E-mail: secretariat@unfccc.int
Web site: http://unfccc.int/2860.php

The 1992 United Nations Framework Convention on Climate Change requires countries to take steps to stabilize greenhouse gas emissions at levels that prevent dangerous anthropogenic interference with the climate system. The Convention calls for industrialized countries to take a leading role in these efforts, including initially stabilizing greenhouse emissions by the year 2000—a goal that almost no countries actually achieved. The

Convention also requires parties to gather and share information on greenhouse gas emissions, national climate policies, and best practices; to launch national strategies for addressing greenhouse gas emissions and adapting to expected impacts; to cooperate in preparing for adaptation to the impacts of climate change; to provide financial and technological support to developing countries; and to report on their efforts. Overall, the UNFCCC is designed to serve as a framework to which specific protocols requiring more binding measures concerning emission reductions can be attached (such as the Kyoto Protocol). The UNFCCC entered into force on March 21, 1994. As of August 2008, there were 192 parties to the Convention.

The UNFCCC Web site contains all official documents relating to the Convention, including the Convention itself, the Kyoto Protocol, decisions of the Conference of the Parties and subsidiary bodies, and much more. In addition, an Essential Background section provides plain-language summaries of the science and the political processes that produced the UNFCCC and Kyoto Protocol. This site is an excellent resource for learning about the development and future of the international climate change regime.

Intergovernmental Organizations

Intergovernmental or international organizations are institutions that have countries as their members. States create intergovernmental organizations to manage their relations on particular issues or in pursuit of particular goals. A large number of international organizations work on issues related to climate change. These include general-purpose UN agencies such as the United Nations Environment Programme and the United Nations Development Programme, multilateral lending institutions such as the World Bank and the regional development banks, and climate-specific entities such as the Intergovernmental Panel on Climate Change. This section provides contact information and very brief descriptions, including the focus of their work on climate change, for a few of the most prominent of these organizations.

African Development Bank (AfDB)
Rue Joseph Anoma
01 BP 1387 Abidjan 01

Cote d'Ivoire
E-mail: afdb@afdb.org
Web site: http://www.afdb.org

AfDB is a multilateral development bank that seeks to improve the welfare of people in Africa and promote policies that lead to economic and social progress, protect public goods, prevent climate change, and improve public health and environmental sustainability as a means of improving socioeconomic conditions. Regional development banks and the larger World Bank play very important roles as multilateral sources of funding for large development projects, including energy projects, and could play pivotal roles in the near future by providing funding for large non-fossil-fuel energy projects.

Alliance of Small Island States (AOSIS)
Division for Sustainable Development
Department of Economic and Social Affairs
Two United Nations Plaza, Room DC2-2220
New York, NY 10017, USA
E-mail: sidsnet@sdnhq.undp.org
Web site: http://www.sidsnet.org/aosis/index.html

Island nations face very serious and increasing threats from sea-level rise caused by global climate change. AOSIS is composed of 43 nations that share these concerns and serves as a voice for small island developing states within the UN system. The AOSIS Web site includes an archive of interventions made by AOSIS members at various international meetings and other documents that provide information about the climate change threats facing small island states and the ways that international regimes can help address them.

Arctic Monitoring and Assessment Programme (AMAP)
Strømsveien 96
P.O. Box 8100 Dep.
N-0032 Oslo, Norway
E-mail: amap@amap.no
Web site: http://www.amap.no/

AMAP provides assistance to the governments of eight Arctic countries on the threats posed to the Arctic by environmental pollution. AMAP is also responsible for implementing components of the Arctic Environmental Protection Strategy, one of which is monitoring and assessing the impacts of anthropogenic pollutants in the Arctic environment, including greenhouse gases. Climate change is already dramatically affecting the Arctic, and AMAP's Web site includes reports on the state of the Arctic climate, expert presentations from meetings on the Arctic climate, and other information.

Asia-Pacific Partnership on Clean Development & Climate (AP6)
E-mail: app_asg@state.gov
Web site: http://www.asiapacificpartnership.org/

AP6 is an international, nontreaty agreement between Australia, China, India, Japan, the Republic of Korea, and the United States in which the countries agree to work together and with the private sector to reduce greenhouse gas emissions and promote secure energy policies. The agreement sets no mandatory limits on greenhouse gas emissions. AP6's Web site includes descriptions and work plans of its task forces, which are focused on individual industrial sectors as well as more general public communiqués.

Asian Development Bank (ADB)
P.O. Box 789
0980 Manila, Philippines
E-mail: information@adb.org
Web site: http://www.adb.org

Comprising both regional and nonregional members, ADB provides funding intended to improve the quality of life and reduce poverty in developing nations in Asia. ADB climate-related programs focus on cleaning the environment, curbing greenhouse gas emissions, increasing energy efficiency, and deploying renewable resource technologies.

Food and Agriculture Organization of the United Nations (FAO)
Viale delle Terme di Caracalla
00100 Rome, Italy
E-mail: fao-hq@fao.org
Web site: http://www.fao.org/

FAO leads international efforts to alleviate and prevent hunger and boost agricultural development. Climate change will have profound impacts on the distribution and viability of agriculture across the world. These issues are addressed through the Interdepartmental Working Group on Climate Change that coordinates FAO's program on climate change. The FAO's primary role with regard to climate change is to assist its member states in adapting to the impacts of climate change on agriculture, forests, and fisheries. This work includes the promotion of climate change mitigation strategies, the adaptation of agricultural systems to climate changes, and the reduction of emissions from the agricultural sector. FAO also participates in climate monitoring systems to help reduce vulnerability to the effects of climate variability. A number of papers posted on the FAO Web site address various aspects of the relationship between climate change and agriculture.

Global Environment Facility (GEF)
1818 H Street NW
Washington, DC 20433, USA
E-mail: secretariat@thegef.org
Web site: http://www.gefweb.org/

GEF is a global financial organization that provides funding to developing countries for projects and programs designed to protect the global environment. Unlike the World Bank, GEF funding need not be repaid. Programs supported by the GEF address six global environmental issues: biodiversity, climate change, international waters, land degradation, the ozone layer, and persistent organic pollutants. GEF also has two special funds dedicated to climate change, including one for least developed countries. GEF manages the new Adaptation Fund under the UNFCCC, which is funded by a 2 percent levy on Clean Development Mechanism transactions. These funds are to be used to support adaptation activities in developing countries. Information on the Web site includes reports on links between economic development and climate change, and brochures laying out priority areas for funding to address climate change.

Intergovernmental Panel on Climate Change (IPCC)
IPCC Secretariat, c/o World Meteorological Organization
7bis Avenue de la Paix, C.P. 2300
CH-1211 Geneva 2, Switzerland
E-mail: ipcc-sec@wmo.int
Web site: http://www.ipcc.ch/

Under the auspices of the World Meteorological Organization and the United Nations Environment Programme, the world's governments created the IPCC to provide periodic, comprehensive, and authoritative assessments of key scientific, technical, and socioeconomic information concerning climate change, including the extent, causes, impacts, and options for adaptation and mitigation of climate change. The purpose of the assessments is to provide national and international policy makers with accurate and up-to-date information on climate science. IPCC releases assessment reports at regular intervals on the current state of knowledge on climate change as well as occasional specialized reports on individual topics. IPCC working groups are composed of dozens of leading experts who work together to review the published literature relating to climate change and summarize the state of knowledge in each area. Hundreds of additional scientists review their work. The IPCC Web site is an invaluable source for anyone researching climate change. It contains all of the IPCC assessment reports and special reports, as well as numerous presentations, charts, graphs, and other background information.

International Energy Agency (IEA)
9 rue de la Fédération
75739 Paris Cedex 15, France
E-mail: info@iea.org
Web site: http://www.iea.org/index.asp

IEA serves as an energy policy advisor for its 26 member states with the aim of ensuring that each has affordable, reliable, and clean energy. Environmental protection is one of three mandates that IEA includes when formulating energy policy. Research conducted by IEA on energy efficiency and emission scenarios has been incorporated into international climate change policies. IEA also maintains a database of its member countries' policies and measures to reduce greenhouse gas emissions, as well as databases on energy efficiency and renewable energy policy. Exhaustive energy statistics and policies of IEA member countries can be accessed through IEA's Web site, along with publications on emissions trading, carbon capture and storage, climate policy approaches, and a variety of other topics related to climate change.

International Union for Conservation of Nature (IUCN)
Rue Mauverney 28
Gland 1196, Switzerland
Web site: http://www.iucn.org

IUCN is the world's largest conservation network and includes 83 national governments, 110 government agencies, and more than 10,000 scientists and experts from 181 countries. Within its focus on conservation, IUCN encourages and helps countries to put in place policies that effectively and efficiently combat climate change, including providing information on greenhouse gas reduction techniques, enhancing natural sinks, and sharing adaptation strategies relevant to conservation.

Organisation for Economic Co-operation and Development (OECD)
2, rue André Pascal
F-75775 Paris Cedex 16, France
Web site: http://www.oecd.org/home/

OECD is made up of countries with democratic governments and market economies. It organizes meetings among its member states, and its secretariat produces publications and statistics covering economic and social issues such as trade, education, development, and science. OECD assists countries in implementing effective and efficient policies to address climate change by conducting policy-relevant research and analysis. OECD's Web site provides access to publications on cities and climate change, development and climate change, the costs and benefits of climate change policies, and a number of other areas that are intended to assist member countries in planning for climate change.

Organization of the Petroleum Exporting Countries (OPEC)
Obere Donaustrasse 93
A-1020 Vienna, Austria
Web site: http://www.opec.org/

OPEC membership currently consists of 12 nations whose economies rely upon revenues from oil exports. OPEC participants coordinate their oil production to help stabilize the oil market, maintain a desired price, and ensure that oil consumers receive stable supplies of oil in order to maintain the economic activity that underpins the demand for oil. OPEC has a strong interest in the international climate negotiations because member countries are highly dependent on oil income. Some OPEC members appear to be working against development of stronger global policy. OPEC's Web site includes the organization's statements to the UNFCCC Conferences of the Parties as well as environmental documents.

United Nations Development Programme (UNDP)
One United Nations Plaza
New York, NY 10017, USA
E-mail: http://www.undp.org/comments
Web site: http://www.undp.org/

UNDP is the chief United Nations agency working on development issues. One of the largest UN agencies, UNDP works with countries on solutions to global and national development challenges and, in this regard, connects and coordinates global and national efforts in reaching the Millennium Development Goals. Ensuring environmental sustainability is one of these goals and requires action against global climate change and ozone-layer depletion. UNDP sponsors projects that will help countries address the challenges presented by climate change. UNDP's Web site provides extensive resources to assist developing countries in particular to reduce their vulnerability to climate change through integrated development planning. It also hosts a database of climate change projects being undertaken by UNDP, as well as a library of documents containing newsletters, reports, and briefings relating specifically to climate change adaptation issues.

United Nations Environment Programme (UNEP)
United Nations Avenue, Gigiri
P.O. Box 30552, 00100
Nairobi, Kenya
E-mail: unepinfo@unep.org
Web site: www.unep.org; www.unep.ch

UNEP leads environmental initiatives in the UN system. It is a relatively small UN agency, with a small budget, so it focuses its efforts, in accordance with its mandate, on gathering and disseminating crucial information and catalyzing and promoting international efforts aimed at environmental protection. Among its many climate change–related programs is the UNEP Finance Initiative, a working group of financial institutions interested in understanding and mitigating their impacts on the environment. The GRID-Arendal center, also part of UNEP, provides credible, science-based environmental information to the public and policy makers through a number of different tools. Resources on the UNEP Web site include numerous publications on climate change.

World Bank
1818 H Street NW
Washington, DC 20433, USA
E-mail: pic@worldbank.org
Web site: http://www.worldbank.org/

The World Bank provides financial and technical support to developing countries, including funding for large development and infrastructure projects. Within the World Bank, the Climate Change Team provides resources and knowledge for the bank's participation in international climate change negotiations as well as in its lending practices. The team also advises the GEF on climate change mitigation strategies and coordinates efforts related to climate change vulnerability and adaptation issues. The Web site includes information about the bank's climate change–related activities and publications on the economic impacts of climate change in different parts of the world.

World Meteorological Organization (WMO)
7bis, avenue de la Paix
Case postale No. 2300
CH-1211 Geneva 2, Switzerland
E-mail: wmo@wmo.int
Web site: http://www.wmo.ch/pages/index_en.html

WMO, part of the UN system, is the international organization responsible for collecting and analyzing data on weather, climate, and the atmosphere. WMO sponsors several programs focused on various aspects of climate change, including climate monitoring and data management, climate prediction, and assessments of the impacts of climate change. The scientific information provided by WMO is utilized by decision makers formulating policy on climate change.

National Government Agencies

Every country has government agencies responsible for dealing with issues relating to different aspects of climate change, such as energy, environmental protection, and scientific research. This section lists the governmental bodies responsible for climate change issues in the largest emitters of greenhouse gases: China, the European Union (EU), India, and the United States, with a concentration on the United States. This list includes contact information for the offices most responsible for responding to climate change and a very brief description of how each organization's work relates to climate change. For some listings, two URLs have been included, one for the agency's main Web site and the other for the agency's specific climate change Web site.

China—National Development and Reform Commission

Department of Resource Conservation and Environmental
    Protection
National Coordination Committee on Climate Change
No. 38, Yuetan Nanjie
Beijing 100824, P.R. China
Web site:
    http://www.ccchina.gov.cn/en/Public_Right.asp?class=25

The National Development and Reform Commission (NDRC), a huge agency, creates and implements economic development policy for China. Within the NDRC, the National Coordination Committee on Climate Change (NCCCC) is responsible for policies related to climate change in China and is the single most important agency in China with regard to national and global climate policy. The NCCCC's Web site provides thorough information on China's official positions on climate change, including related laws and regulation, statements and speeches, national communications, and news about domestic actions that address climate change concerns.

European Commission—Office of the Commissioner for the Environment

Environment Directorate General
Information Centre
BU-9 01/11
B-1049 Brussels, Belgium

E-mail: stavros.dimas@ec.europa.eu
Web site: http://ec.europa.eu/environment/index_en.htm

The European Commission (EC) creates common environmental policies for the 27 member states of the EU. The Commission regulates the environmental quality standards for air and water and controls processes that contribute to air pollution. In addition, it runs the European Climate Change Programme, which helps EU nations reach Kyoto Protocol targets. The Commission was also responsible for creating the EU Emissions Trading Scheme, the largest greenhouse gas trading system in the world. The Environment Directorate Web site includes official documentation of all legislation concerning the environment, as well as a listing of environmental protection activities being undertaken by the EC and member countries. The Climate Action section of the site contains all policy documents, reports, and other materials related to climate change.

India—Ministry of Environment and Forests
CGO Complex, Lodhi Road
New Delhi 110 003, India
E-mail: secymenf@nic.in
Web site: http://envfor.nic.in/

The Ministry of Environment and Forests is the Indian government agency responsible for climate change policies. The agency is committed to making sure that India contributes to the objectives of the UNFCCC. Recently initiated policies in the country include greater use of renewable energy technologies and the improvement of air quality in major cities through the use of clean vehicles, some of which are powered by natural gas.

U.S. Climate Change Science Program (CCSP)
1717 Pennsylvania Avenue NW, Suite 250
Washington, DC 20006, USA
E-mail: information@climatescience.gov
Web site: http://www.climatescience.gov/

This program integrates federal research on climate change and global change. Sponsored by 13 federal agencies, including the Environmental Protection Agency and the Department of Defense, CCSP is overseen by the Office of Science and Technology Policy, the Council on Environmental Quality, the National Economic

Council, and the Office of Management and Budget. CCSP's Web site provides information about the organization of the program, and a library containing synthesis and assessment documents about a host of climate change issues aimed at policy makers and the public as well as many other reports and publications.

U.S. Department of Agriculture (USDA)
1400 Independence Avenue SW
Washington, DC 20250, USA
Web site: http://www.usda.gov/oce/global_change/index.htm

USDA oversees implementation of agricultural policy in the United States. As noted, scientific knowledge continues to increase regarding the ability of soils to release and store carbon, making agricultural policies increasingly relevant to discussions of climate change. The Global Change Program within USDA supports research investigating the roles that terrestrial systems play in influencing climate change and the potential effects of climate change on agricultural and forest systems. Additional USDA research focuses on developing management systems to maintain and enhance food, fiber, and forestry production in the context of a changing climate.

U.S. Department of Energy (DOE)
1000 Independence Avenue SW
Washington, DC 20585, USA
E-mail: the.secretary@hq.doe.gov
Web site: http://www.doe.gov/
Web site: http://www.pi.energy.gov/default.html

DOE is the government agency responsible for U.S. energy policy. Energy policy is critical to addressing climate change because the vast majority of CO_2 emissions result from the burning of fossil fuels for energy. DOE houses the only federal greenhouse gas registry in the United States, a voluntary system that encourages businesses to track and report their emissions. The registry and other climate change programs are administered by DOE's Office of Policy and International Affairs (PI). The PI Web site contains information about the registry and other programs, which include the Clean Energy Initiative; a partnership with companies to reduce greenhouse gas emissions called Climate VISION; and the Climate Change Technology Program, a cross-agency group that works on advancing energy technologies.

U.S. Department of State
Bureau of Oceans and International Environmental and Scientific
    Affairs
2201 C Street NW
Washington, DC 20520, USA
Web site: http://www.state.gov/g/oes/climate/

The State Department manages foreign affairs for the United
States. The department's Bureau of Oceans and International Environmental and Scientific Affairs coordinates U.S. participation
in international environmental negotiations, including meetings
of the UNFCCC. The department also has a leadership role in the
Asia-Pacific Partnership on Clean Development & Climate, along
with other federal agencies, including the Department of Commerce, and carries out other programs related to international cooperation on environmental issues. The Web site provides official
U.S. government positions on climate change policy as presented
at international negotiations. In addition, remarks, briefings, reports, and press releases conveying the official U.S. government
position are archived on the site.

U.S. Environmental Protection Agency (EPA)
Ariel Rios Building
1200 Pennsylvania Avenue NW
Washington, DC 20460, USA
Web site: http://www.epa.gov/climatechange

EPA is the U.S. government agency responsible for protecting the
environment and human health through the development and
enforcement of environmental regulations. With regard to climate, EPA publishes an annual inventory of greenhouse gas
sources and sinks and hosts Climate Leaders, a voluntary emissions reporting program for businesses. In addition, EPA engages
in capacity building to help developing countries reduce emissions of methane, a potent greenhouse gas, and will be heavily involved in future regulation of carbon dioxide emissions. EPA's
regional offices also work on a variety of important air pollution
and other issues relevant to climate change.

U.S. Geological Survey (USGS)
USGS National Center
12201 Sunrise Valley Drive
Reston, VA 20192, USA

Web site: http://www.usgs.gov/

The USGS, a bureau of the U.S. Department of the Interior, studies the landscape of the United States, its natural resources, and the natural hazards that threaten it. Relative to climate change, research supported by the USGS focuses on the interactions between climate, Earth surface processes, and ecosystems. The USGS Global Change Science Program conducts research in a wide range of climate change topics. Its Web site includes data and reports from research projects on sea-level rise, the carbon cycle, climate impacts on species, and many others.

U.S. Global Change Research Information Office (GCRIO)
For contact information, see CCSP
Web site: http://www.gcrio.org/

GCRIO provides access to data and information on climate change research, adaptation and mitigation strategies, and global change–related educational resources. The office also coordinates publications of this information on behalf of federal agencies involved with the USGCRP (see below).

U.S. Global Change Research Program (USGCRP)
For contact information, see CCSP
E-mail: information@usgcrp.gov
Web site: http://www.usgcrp.gov/

USGCRP supports research on the interactions of natural and human-induced changes in the global environment and their implications for society and coordinates the work of the Climate Change Science Program. Research has included work on greenhouse gases, climate change modeling and monitoring, and global carbon and water cycles. Participants in the program include the National Science Foundation, the Smithsonian Institution, and the Department of Transportation, along with many other federal agencies.

U.S. National Aeronautics and Space Administration (NASA)
Public Communications and Inquiries Management Office
NASA Headquarters, Suite 5K39
Washington, DC 20546-0001, USA
E-mail: public-inquiries@hq.nasa.gov
Web site: http://www.giss.nasa.gov

The Earth science program at NASA seeks to better understand Earth's natural systems and apply Earth system science to improve the prediction of climate, weather, and natural hazards. NASA satellite equipment plays an integral role in monitoring changes in the climate system. The major NASA units involved in studying climate change are the Goddard Institute for Space Studies (GISS) and the Goddard Space Flight Center. The GISS Web site contains information on and descriptions of GISS areas of study, data from many different kinds of climate models and other projects, up-to-date maps of temperature trends, and scientific papers published by GISS researchers.

U.S. National Oceanic Atmospheric Administration (NOAA)
14th Street and Constitution Avenue NW, Room 6217
Washington, DC 20230, USA
Web site: http://www.noaa.gov/

NOAA, a scientific agency within the U.S. Department of Commerce, focuses on the conditions of the oceans and atmosphere. On climate change, NOAA seeks to understand and describe climate variability and change to enhance society's ability to plan and respond. NOAA controls several climate observation networks and releases climate forecast predictions. In addition, it researches the impacts climate change is having on natural systems.

Research Institutions

This section lists some of the leading institutions that conduct research on climate science or policy, including contact information and a very brief description of each organization's work on climate change. Thousands of such centers exist, so this list is by no means exhaustive, but it does include many of the most active and prestigious institutions in the field.

Center for International Earth Science Information Network (CIESIN)
Columbia University
Lamont Campus
61 Route 9W
Palisades, NY 10964, USA
E-mail: ciesin.info@ciesin.columbia.edu
Web site: http://www.ciesin.columbia.edu/

CIESIN provides information and analysis to scientists, decision makers, and the public to help them better understand the changing relationship between human beings and the environment. By applying state-of-the-art information technology, CIESIN's work illuminates the geographical distribution of climate vulnerability and the potential impacts of climate change on the world food supply, among other topics. Databases, maps, research findings, and other materials relating to these topics are accessible through the CIESIN Web site.

European Center for Mid-Range Weather Forecasting (ECMWF)
Shinfield Park, Reading
RG2 9A United Kingdom
E-mail: ecmwf-director@ecmwf.int
Web site: http://www.ecmwf.int/

ECMWF provides medium-range weather forecast support to European meteorological organizations and also manages a supercomputer facility that provides resources for weather-forecasting research and computer modeling of the global atmosphere and oceans. The forecasts and research conducted by ECMWF are useful tools for improving the scientific understanding and monitoring of global climate change.

Institute for Global Environmental Strategies (IGES)
2108–11 Kamiyamaguchi
Hayama, Kanagawa
240-0115 Japan
E-mail: www@cola.iges.org
Web site: http://www.iges.or.jp/en/

Established in 1998 by the Japanese government, IGES conducts policy research to support sustainability in countries in the Asia-Pacific region. IGES collaborates with governments and other stakeholders in the region to develop effective policies for mitigation and adaptation to climate change that also facilitate the achievement of regional goals for sustainable development.

International Institute for Applied Systems Analysis (IIASA)
Schlossplatz 1 A-2361
Laxenburg, Austria
Web site: http://www.iiasa.ac.at/

IIASA is an international research agency that explores the impacts that global climate change will have on environmental, economic, technological, social, and human systems. Research at IIASA has looked at land-use changes caused by climate change and the repercussions of climate change for human populations, human health, ecosystems, and energy systems. IIASA also maintains a number of cross-cutting activities, including its Greenhouse Gas Initiative, which is designed to bridge the temporal and spatial scales of the climate change problem.

International Institute for Sustainable Development (IISD)
161 Portage Avenue East, 6th floor
Winnipeg, Manitoba
R3B 0Y4 Canada
E-mail: info@iisd.ca
Web site: http://www.iisd.org

IISD conducts policy research on issues related to sustainable development. In addition to sponsoring and publishing its own research, IISD publishes the *Earth Negotiations Bulletin*, which provides comprehensive daily coverage of international negotiations on treaties pertaining to environmental issues and sustainable development. The outstanding IISD Reporting Services Web site, known as "Linkages" (http://www.iisd.ca/), is home to this bulletin and other environmental newsletters. It contains a large variety of information on many environmental issues, including climate change, provides links to numerous Web sites, and manages electronic mailing lists that individuals can join.

International Research Institute for Climate and Society (IRI)
Columbia University
Lamont Campus
61 Route 9W, Monell Building
Palisades, NY 10964-8000, USA
Web site: http://iri.columbia.edu/

IRI is the world's leading research center focused on improving society's ability to understand, anticipate, and manage climate risk, especially in developing countries. IRI pursues this goal through a number of different strategies, including strategic and applied research on El Niño and long-term climate change, research on climate impacts and adaptation strategies, education, capacity building, forecasts, and information.

Lamont-Doherty Earth Observatory (LDEO)
Columbia University
Lamont Campus
61 Route 9W, Monell Building
Palisades, NY 10964-8000, USA
Web site: http://www.ldeo.columbia.edu

LDEO is one of the world's leading Earth science research institutions, with more than 300 scientists seeking fundamental knowledge about the origin, evolution, and future of the natural world. Part of the Earth Institute at Columbia University, LDEO conducts research on many aspects of climate change, from understanding ancient climates to creating complex computer models to project future climates.

Met Office Hadley Center for Climate Change
FitzRoy Road
Exeter, Devon, EX1 3PB
United Kingdom
E-mail: enquiries@metoffice.gov.uk
Web site: http://www.metoffice.gov.uk/research/hadleycentre/
    index.html

The Met Office is the government meteorological service for the United Kingdom and houses the Hadley Center, a climate science and policy think tank that conducts research and provides input to policy makers in the United Kingdom and worldwide. The Hadley Center is one of the world's preeminent institutions conducting climate modeling and engages in many activities to enable society to understand and adapt to climate change.

Nansen Environmental and Remote Sensing Center
Thormoe hlensgt 47
N-5006 Bergen, Norway
E-mail: admin@nersc.no
Web site: http://www.nersc.no/

The Nansen Center, a research institute affiliated with the University of Bergen, conducts research designed to integrate the use of remote sensing and field observations with numerical modeling in pursuit of improved understanding, monitoring, and forecasting of the global climate system. Specific research projects include studies of sea-ice processes, coastal-zone processes, ecosystem dynamics, and impacts of climate change on the insurance and fishing industries.

National Commission on Energy Policy
1225 I Street, NW, Suite 1000
Washington, DC 20005, USA
E-mail: info@energycommission.org
Web site: http://www.energycommission.org/

The National Commission on Energy Policy explores responses to the rapidly changing landscape of energy needs, vulnerabilities, and opportunities. Drawing on the senior ranks of industry, government, academia, labor, consumer groups, and environmental protection organizations, the commission is a bipartisan group of 20 of the nation's leading energy experts.

Resources for the Future (RFF)
1616 P Street NW
Washington, DC 20036, USA
E-mail: info@rff.org
Web site: http://www.rff.org/

RFF is a nonprofit and nonpartisan organization that conducts independent research—rooted primarily in economics and other social sciences—on environmental, energy, and natural resource issues, including climate change. RFF has helped to advance the application of economics as a tool to develop more environmentally effective policy about the use and conservation of natural resources.

Scripps Institution of Oceanography (SIO)
8602 La Jolla Shores Drive
La Jolla, CA 92037, USA
E-mail: scrippsnews@ucsd.edu
Web site: http://sio.ucsd.edu/

The Scripps Institution of Oceanography is one of the world's preeminent centers for ocean and Earth research, teaching, and public education. A graduate school of the University of California, San Diego, Scripps supports research that focuses on the atmosphere and global climate change, with specific projects covering climate modeling, El Niño, and the impacts of human activities on climate. In addition to these research projects, Scripps offers numerous educational opportunities for both students and teachers interested in the marine and Earth sciences.

Stanford University Global Climate and Energy Project (GCEP)
Peterson Laboratory, Building 550, Room 556Q
416 Escondido Mall
Stanford, CA 94305-2205, USA
E-mail: gcep@stanford.edu
Web site: http://gcep.stanford.edu/

GCEP conducts research and develops technologies that have the potential to provide adequate energy for continued global economic development and poverty alleviation without destabilizing the global climate. Funded by Exxon Mobil, General Electric, Schlumberger, and Toyota, GCEP focuses on technologies with the potential to be effective at a large enough scale to make a substantial contribution to this goal, including solar energy, biomass, hydrogen, and CO_2 capture and storage.

Stockholm Environmental Institute (SEI)
Kräftriket 2B
SE-106 91 Stockholm, Sweden
Web site: http://www.sei.se/

The Stockholm Environmental Institute is an international research facility that specializes in sustainable development and environmental issues. Specific programs at the SEI are focused on the atmospheric environment, energy, and climate change. The SEI publishes reports and provides guidelines for the development of climate change mitigation and adaptation policies.

Tyndall Centre for Climate Change Research
Zuckerman Institute for Connective Environmental Research
School of Environmental Sciences
University of East Anglia
Norwich NR4 7TJ, United Kingdom
E-mail: tyndall@uea.ac.uk
Web site: http://www.tyndall.ac.uk/index.shtml

The Tyndall Centre is a research institution that brings together scientists, economists, social scientists, and engineers who work together to understand and develop sustainable responses to climate change. Research conducted at the center informs climate change policy discussions at the national and international levels.

Woods Hole Oceanographic Institution
266 Woods Hole Road
Woods Hole, MA 02543-1050, USA
E-mail: information@whoi.edu
Web site: http://www.whoi.edu/

The largest nonprofit oceanographic institute in the world, Woods Hole is dedicated to research and higher education in ocean science. A number of projects focus on the relationship between oceans and global climate change. Specific areas of interest include the role of oceans in global warming and abrupt climate change and the development of new systems that monitor the impact of climate change on the oceans.

World Resources Institute (WRI)
10 G Street NE, Suite 800
Washington, DC 20002, USA
E-mail: rspeight@wri.org
Web site: http://www.wri.org

WRI provides nonpartisan, expert guidance on climate change policy and other sustainable development issues to decision makers from governments, international institutions, and the private sector. WRI aims to produce research and analysis that is both scientifically sound and politically practical. WRI is one of the most active and respected NGOs in the field of climate change and has introduced many important policy analyses and technical tools.

Environmental Nongovernmental Organizations

Environmental nongovernmental organizations (NGOs) are independent organizations not beholden to governments or profit-making institutions that work for the protection of the environment. This section provides contact information and a brief description for some of the many NGOs that focus on climate change or related issues such as energy efficiency. It is by no means exhaustive (thousands of NGOs are in operation around the world) but includes many of the groups most active in this area, including both large, general purpose environmental NGOs (such as the Environmental Defense Fund, Greenpeace, and the Natural Resources Defense Council) and climate-specific groups (such as the Climate Action Network). Far more extensive lists can be found on the Internet at sites such as http://library.duke .edu/research/subject/guides/ngo_guide/ngo_links/environ-ment.html.

Alliance to Save Energy
1850 M Street NW, Suite 600
Washington, DC 20036, USA
E-mail: info@ase.org
Web site: http://www.ase.org/

A nonprofit coalition of business, government, environmental, and consumer leaders, the Alliance to Save Energy promotes energy efficiency as a way to build stronger economies and clean the environment. The group advocates for energy-efficient policies that limit greenhouse gas emissions.

Climate Action Network (CAN)
International Secretariat
Charles de Gaulle Str., 5
53113 Bonn, Germany
E-mail: info@climatenetwork.org
Web site: http://www.climatenetwork.org/

CAN is a worldwide network of more than 365 nongovernmental organizations working to promote individual and government action to reduce the impacts of climate change. CAN promotes a three-track approach to mitigating climate change, which includes the use of legal doctrines and sustainable technologies and focuses on the most vulnerable countries. CAN is a very active participant at international climate negotiations and one of the most prominent NGOs working on climate change.

Climate Science Watch (CSW)
Government Accountability Project
1612 K Street NW, Suite 1100
Washington, DC 20006, USA

E-mail: director@climatesciencewatch.org
Web site: http://www.climatesciencewatch.org

Climate Science Watch is a nonprofit interest, education, and advocacy project dedicated to holding public officials accountable with regard to climate science and policy making. CSW's goal is to enable society to respond effectively to global warming and climate change challenges. It is sponsored by the public interest group Government Accountability Project.

Environmental Defense Fund (EDF)
257 Park Avenue South
New York, NY 10010, USA
Web site: http://www.environmentaldefense.org/

The Environmental Defense Fund brings together experts in science, law, policy, and economics to address complex environmental issues and develop solutions to these problems. The Climate and Air team at Environmental Defense Fund pursues strategies to reduce greenhouse gas emissions, promotes the development and use of sustainable energy technologies, and participates in the debate on how to structure government policy on climate change.

Foundation for International Environmental Law and Development (FIELD)
3 Endsleigh Street
London WC1H 0DD, United Kingdom
E-mail: field@field.org.uk
Web site: http://www.field.org.uk/contact.php

FIELD works to advance a fair, effective, and accessible international legal system that will protect the global environment and promote sustainable development. Comprising lawyers from around the world, FIELD works to develop and disseminate law and to actualize it through advocacy, advice, and assistance. With regard to climate change, FIELD has offered legal assistance to the members of the Alliance of Small Island States engaged in UNFCCC negotiations; it has also promoted the implementation of the UNFCCC and the Kyoto Protocol in the European Union.

Friends of the Earth (FOE)
1717 Massachusetts Avenue NW, Suite 600

Washington, DC 20036-2002, USA
E-mail: foe@foe.org
Web site: http://www.foe.org/

Friends of the Earth is an international network of environmental organizations in 70 countries. Member groups both work independently in their home countries and collaborate internationally under the umbrella organization Friends of the Earth International. Active in the fight against climate change, FOE advocates government and private-sector programs to reduce greenhouse gas emissions, protect sinks, and improve energy efficiency.

Global Village Energy Partnership (GVEP)
83 Victoria Street
London SW1H 0HW, United Kingdom
E-mail: info@gvep.org
Web site: http://www.gvep.org/

GVEP works to help developing countries in Latin America, Africa, and Asia set up sustainable energy systems that provide clean energy, reduce poverty, and allow them to reach UN development goals. GVEP was founded at the World Summit on Sustainable Development in Johannesburg in 2002. It collaborates with other energy-related partnerships founded at this meeting.

Greenpeace International
702 H Street NW
Washington, DC 20001, USA
E-mail: info@wdc.greenpeace.org
Web site: http://www.greenpeace.org/

Greenpeace is an international grassroots activist organization that strives to draw attention to environmental problems and to promote action to protect the global environment. With operations around the world, Greenpeace advocates for greater government action and uses public participation to draw attention to issues relating to climate change.

Natural Resources Defense Council (NRDC)
40 West 20th Street
New York, NY 10011, USA
E-mail: nrdcinfo@nrdc.org
Web site: http://www.nrdc.org/

NRDC is an American NGO that uses law, science, and the support of its members and online activists to further its efforts to protect the planet's wildlife and wilderness, ensuring a safe and healthy environment for all living things. NRDC works on many issues, and climate change is one of the major foci of its research and political engagement activities.

Nature Conservancy
4245 North Fairfax Drive, Suite 100
Arlington, VA 22203-1606, USA
Web site: http://www.nature.org

The Nature Conservancy works to protect ecologically important areas around the world. The Conservancy's Global Climate Change Initiative advocates for policies to lower greenhouse gas emissions, works directly to reduce emissions from deforestation, and strives to help natural areas adapt to the impacts of climate change, especially in order to prevent extinctions.

Pew Center on Global Climate Change
2101 Wilson Boulevard, Suite 550
Arlington, VA 22201, USA
Web site: http://www.pewclimate.org/

The Pew Center is a nongovernmental organization that analyzes climate change issues, informs policy makers, works with the business community, and educates the public about climate change and its possible solutions. It also works with government and the private sector to promote the development and implementation of pragmatic policies and practices to reduce greenhouse gas emissions and the risks of dangerous climate change.

Union of Concerned Scientists (UCS)
2 Brattle Square
Cambridge, MA 02238-9105, USA
Web site: http://www.ucsusa.org

UCS is a leading science-based nonprofit organization working on environmental, security, technological, and other issues. UCS produces independent research and analysis that it uses to inform the public and policy makers on issues such as climate change and the policy steps needed to address it, including emissions reductions. UCS also actively engages with the federal legislative process.

World Wildlife Fund/Worldwide Fund for Nature (WWF)
1250 Twenty-fourth Street NW
P.O. Box 97180
Washington, DC 20090-7180, USA
Web site: http://www.worldwildlife.org

Known as the World Wildlife Fund in the United States and the Worldwide Fund for Nature elsewhere in the world, WWF is a multinational conservation organization that works on many environmental issues, including climate change. The main areas in which the WWF works to combat climate change are the creation of effective policies, business emissions reduction, adaptation strategies for high-risk communities, and the protection of forests that mitigate climate change.

Business-Oriented International Nongovernmental Organizations

The following section lists some of the most prominent business-affiliated groups that engage in climate change issues. Each listing provides contact information and a brief description of the organization's work on climate change. This list concentrates on organizations that focus specifically on climate change and related issues, as opposed to general industry associations (groups made up of companies from specific industries, such as the World Coal Institute or the International Council on Mining and Metals), even though many of these areas are also active on climate-related issues.

American Council on Renewable Energy (ACORE)
1629 K Street NW, Suite 210
Washington, DC 20033-3518, USA
Web site: http://www.acore.org/

ACORE is a member organization that includes renewable energy industries and associations, utilities, and others interested in working toward greater deployment of renewable energy technologies in the United States. ACORE provides a forum for its members to

learn about developments in renewable energy and pool their resources and influence to promote it. Annual meetings and publications such as informational guides provide greater knowledge of
the benefits of using renewable energy sources.

Business Council for Sustainable Energy (BCSE)

1620 Eye Street NW, Suite 501
Washington, DC 20006, USA
E-mail: http://www.acore.org/contact/
Web site: http://www.bcse.org/

BCSE seeks a business-oriented approach to environmental and
energy-related issues. Comprising leading businesses from various sectors that produce or use energy, BCSE promotes the use of
market-based instruments to reduce pollution and the development of renewable energy resources. The U.S. office of BCSE
works with sister organizations in the United Kingdom, EU, and
Australia to promote sustainable energy policy and competitive
business environments worldwide.

Carbon Disclosure Project (CDP)

40 Bowling Green Lane
London EC1R 0NE, United Kingdom
E-mail: info@cdproject.net
Web site: http://www.cdproject.net/

CDP provides guidance to investors concerned about the impacts
of climate issues on the value of their investments and creates a
forum for companies taking decisive action to mitigate their carbon risk. CDP sends out an annual questionnaire requesting information regarding companies' response to climate change; the
questionnaire also includes a request for the disclosure of total
greenhouse gas emissions. This information is used to create
widely read annual reports, to facilitate a dialogue, and to spur
the creation of rational business responses to climate change.

Ceres

99 Chauncy Street, 6th floor
Boston, MA 02111, USA
Web site: http://www.ceres.org/

Ceres is a national network of investors, environmental organizations, and other public interest groups working with companies

and investors to address sustainability challenges such as global climate change. It coordinates the Investor Network on Climate Risk, works to promote corporate carbon-risk disclosure, and conducts research on a variety of climate-related topics of interest to business and investors.

The Climate Group
The Tower Building, 3rd floor
York Road
London SE1 7NX, United Kingdom
E-mail: info@theclimategroup.org
Web site: http://www.theclimategroup.org/

The Climate Group is an independent, nonprofit organization dedicated to advancing business and government leadership on climate change. The group encourages the development and sharing of expertise on how its members can contribute to creating a low-carbon economy while increasing global prosperity.

Combat Climate Change (3C)
c/o Vattenfall AB
SE-162 87 Stockholm, Sweden
Web site: http://www.combatclimatechange.org/

The stated goal of the 3C Initiative is to underscore the need for urgent action by the global community and to influence the current climate regime by demanding a global framework supporting a market-based solution to climate change. Its main activity is collecting signatures for the 3C Statement, which outlines nine principles that promote its road map to a low-emitting society.

Corporate Leaders Group on Climate Change (CLG)
University of Cambridge Programme for Industry
1 Trumpington Street
Cambridge CB2 1QA, United Kingdom
E-mail: info@cpi.cam.ac.uk
Web site: http://www.cpi.cam.ac.uk/bep/clgcc/

This group of business leaders works in partnership with the UK government to strengthen domestic and international progress on reducing greenhouse gas emissions. CLG advocates ambitious action to stimulate low-carbon technology and the adoption of challenging long-term emissions-reduction targets.

e5: European Business Council for Sustainable Energy
Hauptstrasse 43
D-61184 Karben, Germany
E-mail: office@e5.org
Web site: http://www.e5.org

A council of European business leaders concerned about the future of the economy and the environment, e5 acknowledges that companies in industrialized countries should take particular responsibility and advocates the adoption of progressive energy, transport, and related policies to avoid ecological risks and long-term economic damage from climate change. e5 acts as a proactive representative of its members' interests in EU energy policy debates and international climate negotiations.

European Climate Forum (ECF)
c/o PIK Potsdam
Telegrafenberg A31
D-14473 Potsdam, Germany
E-mail: info@european-climate-forum.net
Web site: http://ecf.pik-potsdam.de/

ECF is a platform for joint studies and science-based stakeholder dialogues on climate change. It attempts to facilitate dialogue that will clarify differences in stakeholder opinion and to produce analyses that summarize and advance the state of knowledge in critical problem areas relating to climate policy.

Gleneagles Dialogue Industry Partnership Project
World Economic Forum
91–93 route de la Capite
CH-1223 Cologny/Geneva, Switzerland
E-mail: contact@weforum.org
Web site: http://www.weforum.org/

The Gleneagles Dialogue Industry Partnership Project is a high-level public-private partnership project open to the industry partners of the World Economic Forum and focused on climate change. The activities of the group include engaging the private sector with the Group of Eight/Group of Twenty governments across three key areas of climate change: policy frameworks, smarter investment, and consumption behavior.

Global Roundtable on Climate Change (GROCC)
The Earth Institute at Columbia University
B-100 Hogan Hall 3277
2910 Broadway
New York, NY 10025, USA
E-mail: grocc@ei.columbia.edu
Web site: http://www.earthinstitute.columbia.edu/grocc/

GROCC brings together critical high-level stakeholders—senior executives from the private sector and leaders of international governmental and nongovernmental organizations—from all regions of the world together with leading scientists and technology experts to discuss and explore areas of potential consensus regarding core scientific, technological, and economic issues critical to shaping sound public policies on climate change.

Institutional Investors Group on Climate Change (IIGCC)
c/o The Climate Group
The Tower Building, 3rd floor
York Road
London SE1 7NX, United Kingdom
E-mail: stephanie.pfeifer@iigcc.org
Web site: http://www.iigcc.org

IIGCC is a forum for collaboration between pension funds and other institutional investors on issues related to climate change. IIGCC seeks to promote a better understanding of the implications of climate change on investors and encourages investors to advocate for a low-carbon economy in the markets in which they invest.

International Chamber of Commerce (ICC)
38, Cours Albert 1er
75008 Paris, France
Web site: http://www.iccwbo.org/

The International Chamber of Commerce, a corporate-member supported organization, promotes open economic systems and the global economy as forces for economic growth, job creation, and prosperity. The chamber's Commission on the Environment and Energy develops its positions on environmental issues, including climate change. ICC is recognized by the UNFCCC process as the lead representative of business in international climate negotiations.

United Kingdom Business Council on Sustainable Energy (UKBCSE)
35/37 Grosvenor Gardens
London SW1W 0BS, United Kingdom
E-mail: info@bcse.org.uk
Web site: http://www.bcse.org.uk/

The UKBCSE brings together businesses, government, and other high-level stakeholders to discuss policy issues and facilitate the transition to greater use of renewable fuels and other clean-energy technologies. Along with facilitating this dialogue, the UKBCSE highlights the scope for technological and market innovation to deliver key environmental goals. The group also promotes policies to sustain long-term market growth.

World Business Council for Sustainable Development (WBCSD)
4 Chemin de Conches
1231 Conches-Geneva, Switzerland
E-mail: info@wbcsd.org
Web site: http://www.wbcsd.org/

Composed of leaders from more than 200 global businesses, WBCSD works to elaborate and promote the business case for sustainable development. Together with WRI, WBCSD developed the Greenhouse Gas Protocol Initiative to harmonize international greenhouse gas accounting and reporting standards and help enable businesses to more easily track and report their emissions. The group also strongly promotes energy efficiency as a way to encourage sustainable development in business and reduce greenhouse gas emissions.

8

Resources

This chapter provides an annotated bibliography of selected books, articles, and reports, as well as Internet, film, and video resources about climate science and policy. It is divided into several sections: books on climate science, books on climate policy and economics, books for young adults, books on ways to get more involved, academic journals, Internet resources, and film and video.

A vast quantity of information is being produced on the subject of climate change. Since we cannot include all the relevant publications, we have concentrated on materials written for a general audience. More technical sources are included only if they are regarded as critically important. Moreover, because the field is developing so quickly, we have only included materials published after 2005, with a few exceptions for particularly relevant works. Naturally, this list is still just a starting point for further reading, not a comprehensive list.

Books

Climate Science

Bigg, G. *The Oceans and Climate*. Cambridge, UK: Cambridge University Press, 2006, 286 pp. ISBN-13: 978-0521016346

The Oceans and Climate is a textbook intended for undergraduate and graduate students studying Earth and environmental sciences, oceanography, meteorology, and climatology. Chapters examine the relationship and interactions between oceans and climate in terms of the past, the present, and the future and discuss ocean changes linked to climate over a variety of time scales. Additional sections of the book discuss the physical, chemical, and biological interactions between the atmosphere and the oceans.

Broecker, W. S., and R. Kunzig. *Fixing Climate: What Past Climate Changes Reveal About the Current Threat—and How to Counter It.* **New York: Hill and Wang, 2008, 272 pp. ISBN-13: 978-0809045013**

Fixing Climate presents a history of the scientific enquiry of global warming theory by tracing the story from the 19th century to 1957, the dawn of the modern era of greenhouse studies to 2008. The authors explore how we arrived at the point where climate change is no longer preventable. Nonetheless, they offer a glimmer of hope to the unstoppable warming with developments of new technologies that reduce carbon dioxide output and dispose of it harmlessly.

Calvin, W. H. *Global Fever: How to Treat Climate Change.* **Chicago: University Of Chicago Press, 2008, 352 pp. ISBN-13: 978-0226092041**

Calvin approaches climate change as any doctor would by consulting the climate doctors and looking at the lab results to put together the treatment options and to make the diagnosis. He first clearly sets out the current state of the Earth's warming climate as well as the disastrous possibilities if nothing is done. To avoid this dire fate, *Global Fever* issues a stark warning and an ambitious blueprint for addressing it.

Dessler, A. E., and E. A. Parson. *The Science and Politics of Global Climate Change: A Guide to the Debate.* **Cambridge, UK: Cambridge University Press, 2006, 200 pp. ISBN-13: 978-0521539418**

The authors combine their expert knowledge of the atmosphere and public policy to help clarify the climate change debate and provide a general audience with information on both the science and politics of climate change. The book provides useful background information on how scientific knowledge is developed, how policy debates work, and the complexities of developing public policy that adequately incorporates science. This section is followed by a summary of the current knowledge about global climate change and the options available to address it.

Flannery, T. *The Weather Makers: How Man Is Changing the Climate and What It Means for Life on Earth.* New York: Atlantic Monthly Press, 2006, 384 pp. ISBN-13: 978-0871139351

The Weather Makers is a guide to global warming, atmospheric science, and climate change, written for a broad audience. It describes how the atmosphere and climate system work; the ecological impacts of climate change; and how climate change may impact weather patterns, natural systems, and human life. The author also provides information on how individuals can make a difference by taking actions to mitigate climate change.

Gerard, D., and E. Wilson. *Carbon Capture and Sequestration: Integrating Technology, Monitoring, Regulation.* New York: Blackwell, 2007, 296 pp. ISBN-13: 978-0813802077

This book presents an overview of the technical, legal, and economic aspects of using carbon capture and sequestration to reduce greenhouse gas levels. The first part of the book offers a technological summary of carbon capture and sequestration technology, including sections on how the process works, ways to monitor it, and the risks involved with using it. The second part discusses how policies can be formulated to encourage the use of this technology, including an overview of relevant legal, economic, and political issues.

Goodstein, E. *Fighting for Love in the Century of Extinction: How Passion and Politics Can Stop Global Warming.* Boston: University Press of New England, 2008, 164 pp. ISBN-13: 978-1584656579

The central idea in *Fighting for Love in the Century of Extinction* is that unchecked global warming threatens to destroy one of every two animals, birds, plants, reptiles, forests, fish, and other creatures alive today. To ward off this century of mass extinction, Goodstein offers practical political solutions.

Kolbert, E. *Field Notes from a Catastrophe: Man, Nature, and Climate Change.* London: Bloomsbury, 2006, 240 pp. ISBN-13: 978-1596911307

Field Notes from a Catastrophe summarizes the author's firsthand experiences gained while visiting countries where the impacts of global warming are already being felt and provides evidence of climate change from a number of perspectives. This well-written book considers the underlying causes of climate change, its likely impacts, and possible solutions through policy processes and behavior change.

Lynas, M. *Six Degrees: Our Future on a Hotter Planet.* **Washington, DC: National Geographic, 2008, pp. 336. ISBN-13: 978-1426202131**

The Intergovernmental Panel on Climate Change projects average global surface temperatures to rise between 1.4° and 5.8°C by the end of this century. Based on this forecast, Mark Lynas outlines what to expect from a warming world, degree by degree.

Mann, M., and L. Kump. *Dire Predictions: Understanding Global Warming.* **Upper Saddle River, NJ: Prentice Hall, 2008, 120 pp. ISBN-13: 978-0136044352**

Dire Predictions is authored by two esteemed climate scientists who distill vast volumes of data and information about climate change into accessible text and revealing graphics. This volume synthesizes critical findings from across the many academic disciplines involved in the study of climate change to help the general public understand and make behavioral and political decisions related to climate change.

Mooney, C. *Storm World: Hurricanes, Politics and the Battle over Global Warming.* **New York: Harcourt, 2007, 400 pp. ISBN-13: 978-0151012879**

Chris Mooney, a science journalist, wrote this book in response to Hurricane Katrina after it devastated the Gulf Coast in 2005. In it, he explores how global warming may affect hurricanes and traces the history of the science and politics of hurricane predictions.

Pearce, F. *With Speed and Violence: Why Scientists Fear Tipping Points in Climate Change.* **Boston: Beacon Press, 2007, 278 pp. ISBN-13: 978-0807085769**

With Speed and Violence explains why some scientists fear that, without immediate action, the world will face grave climatic consequences in the future. Much of the research presented in the book focuses on past climate change and how scientists are only now learning how fast and violent these changes were and what this understanding may mean for the future. The book, written for nonscientists, provides a complete overview on global warming and the effects it may have on the climate system.

Schlesinger, M. *Human-Induced Climate Change: An Interdisciplinary Assessment.* **Cambridge, UK: Cambridge University Press, 2007, 504 pp. ISBN-13: 978-0521866033**

Human-Induced Climate Change provides a review of climate change science, impacts, mitigation, adaptation, and policy. The book describes recent scientific findings on how the climate system is changing, the impacts that climate change will have on various economic sectors and geographic regions, the current mitigation technologies that exist, and how to design effective climate policy.

Shackley, S., and C. Gough. *Carbon Capture and Its Storage: An Integrated Assessment.* **London: Ashgate, 2006, 313 pp. ISBN-13: 978-0754644996**

Carbon Capture and Its Storage provides an excellent overview of the basics of carbon capture and storage, laws regulating its use, and its public perception. Case studies focusing on sites in the United Kingdom highlight the successes of the use of technology and compare it to other energy alternatives. The book concludes with a list of recommendations on how carbon capture and storage can be used in the future.

Spratt, D., and P. Sutton. *Climate Change Code Red: The Case for Emergency Action.* **Melbourne, Australia: Friends of the Earth, 2008, 256 pp. ISBN-13: 978-1920767082**

Climate Change Code Red makes the case for emergency action by revealing extensive scientific evidence that the global warming crisis is far worse than official reports and national governments have indicated. The book shows that serious climate change impacts are already happening—and at a more rapid pace and at lower global temperature increases than projected. According to Spratt and Sutton, the case for emergency action is not a radical idea but the only course of action guaranteed to protect our planet.

Walker, G., and D. King. *The Hot Topic: What We Can Do About Global Warming.* Orlando, FL: Harvest Books, 2008, 276 pp. ISBN-13: 978-0156033183

Walker, the author of *An Ocean of Air,* and King, the UK's chief science adviser, present a concise guide on the science behind global warming as well as the technological, governmental, and economic resources we have to approach this problem. Also included is an appendix that dispels many myths and misconceptions about climate change.

Weart, S. *The Discovery of Global Warming.* Cambridge, MA: Harvard University Press, 2004, 240 pp. ISBN-13: 978-0674016378

The Discovery of Global Warming provides a comprehensive history of climate change, with an emphasis on the long, multifaceted process that led to the scientific consensus on global warming. The author identifies major players in the field of climate science and describes key discoveries and how those discoveries have reinforced and continue to build upon one another to produce an increasingly nuanced view of the climate system and our impacts on it. Weart looks closely at the difficulties of trying to reach agreements (whether scientific or political) on issues that have a critical impact on human life.

Climate Policy and Economics

Adger, W. N., J. Paavola, S. Huq, and M. J. Mace. *Fairness in Adaptation to Climate Change.* Cambridge, MA: MIT Press, 2006, 312 pp. ISBN-13: 978-0262511933

Fairness in Adaptation to Climate Change examines social justice issues associated with climate change adaptation strategies. The authors argue that policy responses to climate change must not place unfair burdens on developing nations, which are already extremely vulnerable to the changing climate. Drawing from the fields of economics, policy analysis, and environmental science, the book provides an overview of the background of different justice issues applied to climate change and describes both present and future fairness issues of climate change policies, with the use of specific case study analysis from nations around the world.

Burton, I., and N. Leary. *Climate Change Vulnerability and Adaptation.* **2 vols. London: Earthscan, 2008, 1,088 pp. ISBN-13: 978-1844074808**

Climate Change Vulnerability and Adaptation was released in response to the work of the Assessments of Impacts and Adaptations to Climate Change project, launched by the Intergovernmental Panel on Climate Change (IPCC) in 2002. The project was an investigation aimed at understanding climate change vulnerabilities in developing countries and the adaptation strategies that can be used there. Written by experts in climate change vulnerability and adaptation, the book explores how new methods in both these areas can be applied to specific sectors and resources, with the use of case studies from around the world.

Cline, W. R. *Global Warming and Agriculture: End-of-Century Estimates by Country.* **Washington, DC: Peterson Institute, 2007, 250 pp. ISBN-13: 978-0881324037**

In *Global Warming and Agriculture,* Cline details the impacts that global warming will have on the economies of developing nations, which rely heavily on agriculture. The book discusses model simulations of climate change occurring late in the 21st century and the impacts it will have on temperature, precipitation, and, most important, agriculture. Most developing countries rely on agriculture and will be hardest hit by these effects.

Depledge, J. *The Organization of Global Negotiations: Constructing the Climate Change Regime.* **London: Earthscan, 2005, 256 pp. ISBN-13: 978-1844070466**

The Organization of Global Negotiations provides a complete overview of the international negotiations that led to global climate change agreements, including the Kyoto Protocol and the United Nations Framework Convention on Climate Change (UNFCCC), and reveals the difficulties of the international policymaking process. The book discusses the major players in global climate negotiations, where negotiations have taken place, the scientific basis of the agreements, and a review of how agreements have worked up to this point. The book is intended as a guide for those involved in climate change or other international negotiations.

Dow, K., and T. E. Downing. *The Atlas of Climate Change: Mapping the World's Greatest Challenge.* **Berkeley, CA: University of California Press, 2006, 112 pp. ISBN-13: 978-0520250239**

The Atlas of Climate Change uses maps, charts, graphs, and pictures to explore the causes of climate change, the potential impacts it will have on human life and natural systems, and policy options available to reduce these negative consequences. This informative and very accessible book provides a historical overview of climate change, covering past greenhouse gas trends, present-day warning signs, and future scenarios. Additional sections discuss the impacts that climate change may have on human health and society, as well as the policy options available to address it.

Fleming, J. R. *Historical Perspectives on Climate Change.* **New York: Oxford University Press, 2004, 208 pp. ISBN-13: 978-0195078701**

This book provides a historical description of perspectives on climate change from the Enlightenment through the late 20th century; it also shows how past events have influenced the ongoing environmental debate. Based on primary and archival sources, the volume offers insight into what individuals in the past thought, experienced, and understood about climate change.

Gore, Albert, Jr. *An Inconvenient Truth: The Planetary Emergency of Global Warming and What We Can Do About It.* **New York: Rodale, 2006, 328 pp. ISBN-13: 978-1594865671**

An Inconvenient Truth, which shares the same name as the Academy Award–winning documentary with which it was released, is a complete guide to global warming, written by former U.S. Vice President Al Gore. The book is based on a series of global warming presentations given by Gore, who has understood the risks and publicly advocated for strong policy on climate change since the early 1980s. Gore uses the latest research from scientists to describe the state of the climate system and address the hazards that global warming will bring. He argues that global warming is not in question and stresses the need for immediate action to prevent climate change, for without it, the consequences may prove disastrous in the future.

Hoffmann, A., and J. Woody. *Climate Change: What's Your Business Strategy?* **Cambridge, MA: Harvard University Press, 2008, 138 pp. ISBN-13: 978-1422121054**

Hoffman, a professor at the University of Michigan, who consults with corporations about environmental strategies, and Woody, who works at a renewable energy venture capital fund, present advice to help companies assess their carbon emissions and develop ways to reduce them that actually provide the firm with cost savings and other advantages over competitors.

Krupp, F., and M. Horn. *Earth: The Sequel—The Race to Reinvent Energy and Stop Global Warming.* **New York: W. W. Norton, 2008, 256 pp. ISBN-13: 978-0393066906**

Coauthored by Fred Krupp, president of the business-focused environmental nongovernmental organization Environmental Defense Fund, and journalist Miriam Horn, this book argues that the only way for the United States to slow, stop, and reduce its greenhouse gas emissions is to harness the power of capitalism. Specifically, the authors support a mandatory cap on carbon emissions, which would increase the cost of doing business using old energy technologies and induce innovators to come up with carbon-reducing solutions.

Labatt, S., and R. R. White. *Carbon Finance: The Financial Implications of Climate Change.* **Hoboken, NJ: Wiley, 2007, 268 pp. ISBN-13: 978-0471794677**

Carbon Finance provides an overview of carbon finance, specific techniques used by professionals, the carbon emissions market, the impact the Kyoto Protocol has on carbon finance, and different risk-transfer methods used for reducing the risk of weather, including the use of weather derivatives, weather indexes, and catastrophe bonds.

Linden, E. *The Winds of Change: Climate, Weather, and the Destruction of Civilizations.* **New York: Simon & Schuster, 2006, 320 pp. ISBN-13: 978-0684863528**

The Winds of Change is a complete guide to the importance of climate change for human societies; it includes evidence of impacts on early civilization and those associated with present-day climate phenomena. With a thorough explanation of El Niño and case studies such as Hurricane Katrina and the European heat wave of 2003, the book provides a vivid glimpse of how completely dependent our economic systems are on climate and weather and of how future change may affect us.

Menne, B., and K. L. Ebi. *Climate Change and Adaptation Strategies for Human Health.* Darmstadt, Germany: Steinkopff-Verlag, 2006, 449 pp. ISBN-13: 978-3798515918

This book reports the results from the *Climate Change and Adaptation Strategies for Human Health* study, conducted in Europe from 2001 to 2004 by the World Health Organization and the European Union. The study provides examples of how society is not ready to face the health impacts of climate change and advocates a new, proactive approach to managing public health in response to climate change, partly through the use of better prediction methods and early warning systems.

Nordhaus, W. *A Question of Balance.* New Haven, CT: Yale University Press, 2008, 256 pp. ISBN-13: 978-0300137484

William Nordhaus integrates the entire spectrum of economic and scientific research to weigh the costs of reducing emissions against the benefits of reducing the long-run damages from global warming. The book offers extensive analyses of the economic and environmental dynamics of greenhouse gas emissions and climate change and provides the tools to evaluate alternative approaches to slowing global warming. The author emphasizes the need to establish effective mechanisms, such as carbon taxes, to harness markets and harmonize the efforts of different countries.

Oberthur, S., and H. E. Ott. *The Kyoto Protocol: International Climate Policy for the 21st Century.* New York: Springer, 2006, 359 pp. ISBN-13: 978-3540664703

This book provides an overview of the history of international environmental negotiations, with a specific focus on the debates that led up to the signing of the Kyoto Protocol and the implications that this agreement will have on preventing climate change. It describes early international negotiations on climate change, provides an analysis of the entire Kyoto Protocol, and provides an outlook for future negotiations.

Ruth, M., K. Donaghy, and P. Kirshen. *Regional Climate Change and Variability: Impacts and Responses.* Northampton, MA: Edward Elgar, 2006, 260 pp. ISBN-13: 978-1845425999

Climate change is expected to have negative impacts on human health. *Regional Climate Change and Variability* uses case studies to develop a framework to protect human health from the impacts of global climate change. The cases are used to identify what areas of modern-day health systems must be updated to adapt to climate change and to learn about the processes that were used to implement the policy changes.

Shellenberger, M., and T. Nordhaus. *Break Through: From the Death of Environmentalism to the Politics of Possibility.* **New York: Houghton Mifflin, 2007, 256 pp. ISBN-13: 978-0618658251**

In previous work, Shellenberger and Nordhaus have argued that the methods of the environmental movement will not be satisfactory to slow global warming. In this book, they suggest that the emerging ecological crisis demands harnessing human potential and the power of economic development in new and different ways. Thus, the authors argue for big, bold investments that will pave the way to a new energy economy.

Sjöstedt, G. *Climate Change Negotiations: A Guide to Resolving Disputes and Facilitating Multilateral Cooperation.* **London: Earthscan, 2007, 320 pp. ISBN-13: 978-1844074648**

Climate Change Negotiations suggests solutions to roadblocks in international climate change negotiations from the perspectives of a policy maker, a negotiator, a lawyer, a scientist, and a sociologist. Each provides their views about important unresolved issues and how negotiations can produce an effective agreement. The book provides details on setbacks in international climate negotiations, which include the dominance of developed nations, airline industry emissions, insurance and risk-transfer instruments, and the problems of cost-benefit analysis.

Stern, N. *The Economics of Climate Change: The Stern Review.* **Cambridge, UK: Cambridge University Press, 2007, 712 pp. ISBN-13: 978-0521700801**

The Stern Review is the most comprehensive analysis of the economic impacts of global climate change yet undertaken. A task force led by Sir Nicholas Stern, former head of the UK Government Economic Service and a former chief economist of the World Bank, analyzed the economic impacts of climate change, the economics of stabilizing greenhouse gas emissions, policy responses for adaptation and mitigation, and a call for international collective action. The review concludes that failing to respond will be substantially more expensive than acting immediately to dramatically reduce emissions.

Tidwell, M. *The Ravaging Tide: Strange Weather, Future Katrinas, and the Coming Death of America's Coastal Cities.* **New York: Free Press, 2006, 208 pp. ISBN-13: 978-0743294706**

The Ravaging Tide warns that, without action to fight climate change, future storms and the impacts of global warming will become more severe. The author's main argument places the blame for storms like Hurricane Katrina on the U.S. government and its energy policy. The book explores the impacts that future storms and global warming will have across the world. The author advocates for immediate action, and the book lays out steps to change our energy system to one that is clean and therefore can prevent the dangerous consequences of climate change.

Van Drunen, M. A., R. Lagase, and C. Dorlands. *Climate Change in Developing Countries.* **Oxfordshire, UK: CABI Publishing, 2006, 320 pp. ISBN-13: 978-1845930776**

Climate Change in Developing Countries summarizes the studies conducted by the Netherlands Climate Change Studies Assistance Programme, which was set up to help developing nations fulfill requirements of the UNFCCC. The report evaluates the vulnerability to climate change; mitigation and adaptation strategies; and the climate change policies of Bolivia, Colombia, Ecuador, Egypt, Ghana, Kazakhstan, Mali, Mongolia, Senegal, Suriname, Vietnam, Yemen, and Zimbabwe.

Young Adult

David, L., and C. Gordon. *The Down-to-Earth Guide to Global Warming.* **London: Orchard, 2007, 128 pp. ISBN-13: 978-0439024943**

Written by the producers of Al Gore's documentary *An Inconvenient Truth, The Down-to-Earth Guide* is filled with facts, photos, and illustrations that help kids understand the causes and potential consequences of climate change. The book also has suggestions about ways kids can help fight global warming in their homes, schools, and communities.

Evans, K. *Weird Weather: Everything You Didn't Want to Know about Climate Change But Probably Should Find Out.* Berkeley, CA: Groundwood, 2007, 96 pp. ISBN-10: 0888998384

Weird Weather is a comic book that explores climate change from the perspectives of an idealistic adolescent, fat-cat businessmen, and a mad scientist. The graphic format makes the text accessible to almost anyone interested in learning about climate change. The book features sections on the science behind global warming, how the climate system is presently changing, and what can be done to prevent problems from occurring in the future. Lists of references and suggested further reading are included as well.

Revkin, A. *The North Pole Was Here: Puzzles and Perils at the Top of the World.* New York: Kingfisher, 2006. pp. 128. ISBN-10: 0753459930

In 2003, *New York Times* reporter Andrew Revkin traveled with a group of scientists and researchers into the Arctic, the region at greatest risk from climate change. This book discusses the history of Arctic exploration and the geology of the North Pole, as well as the likely impacts of climate change on the region.

Silver, J. *Global Warming and Climate Change Demystified.* New York: McGraw-Hill Professional, 2008, pp. 289. ISBN-13: 978-0071502405

Global Warming and Climate Change Demystified begins by looking at the causes of both natural and anthropogenic climate change. The book then goes on to examine the consequences and possible solutions that can be implemented. The fundamentals of global warming and climate change are presented in an unbiased and thorough manner. As with all books in the *Demystified* series, there are end-of-chapter quizzes and final reviews to test your knowledge.

Woodward, J. *Climate Change.* DK Eyewitness Books. New York: DK Publishing, 2008, pp. 72. ISBN-13: 978-0756637705

This book provides an in-depth but easily accessible look at the phenomenon of global warming and asks the questions of what's causing it, what might lead to it, and what we can do to fight back.

What Can I Do?

David, L. *Stop Global Warming: The Solution Is You!* **Golden, CO: Fulcrum, 2006, 70 pp. ISBN-13: 978-1555916213**

This resource explains how ordinary people can get involved in the fight against climate change and why taking action is so important. The book presents a brief overview of the basics of climate change and provides several suggestions to the reader on how he or she can help stop global warming. The book also includes quotes from celebrities, musicians, and Americans who have already taken action, which help reinforce the book's main ideas.

De Rothschild, D. *The Live Earth Global Warming Survival Handbook: 77 Essential Skills to Stop Climate Change.* **New York: Rodale Books, 2007, 160 pp. ISBN-13: 978-1594867811**

The official companion to the Live Earth concerts, this survival guide presents scientific and environmental facts and practical advice for stopping climate change and living through it.

Goodall, C. *How to Live a Low-Carbon Life.* **London: Earthscan, 2007, 320 pp. ISBN-13: 978-1844074266**

How to Live a Low-Carbon Life is a comprehensive guide to calculating personal carbon emissions and explains how individuals can reduce their greenhouse gas emissions. The book features easy-to-read reference tables on how to calculate your own carbon footprint and has a companion Web site with resources on the latest environmentally sustainable products and technologies. It is perfect for families looking to learn how to reduce their environmental impact.

Horn, G. *Living Green: A Practical Guide to Simple Sustainability.* **New York: Freedom Press, 2006, 171 pp. ISBN-13: 978-1893910478**

Living Green is a practical guide to living a sustainable lifestyle. The book provides a history of the green movement and discusses simple changes all individuals can make to help save the environment and mitigate climate change. The book covers sustainable health, sustainable homes, and how to build a sustainable future. Also included in the book is a resource and product guide that can be used by individuals in their quest to go green.

Isham, J., and S. Waage. *Ignition: What You Can Do to Fight Global Warming and Spark a Movement.* Washington, DC: Island Press, 2007. ISBN-10: 978-1597261562

This book includes chapters from celebrated writers, renowned scholars, young activists, and advocates who draw on direct experience in grassroots organization, education, law, and social leadership to offer solutions on how to start the ultimate green revolution.

Keller, G., D. Jenks, and J. Papasan. *Green Your Home: The Proven Path to Sustainability.* New York: McGraw-Hill, 2008, 200 pp. ISBN-13: 978-0071489850

This volume offers advice to homeowners who want to increase energy efficiency, reduce waste, and reduce their impact on the planet in general.

Rogers, E. and T. M. Kostigen. *The Green Book: The Everyday Guide to Saving the Planet One Simple Step at a Time.* New York: Three Rivers Press, 2007, 224 pp. ISBN-13: 978-0307381354

Rogers and Kostigen provide better lifestyle choices for 12 arenas, such as home, work, and school. Celebrities such as Ellen DeGeneres, Robert Redford, Will Ferrell, Jennifer Aniston, Faith Hill, Tyra Banks, Dale Earnhardt Jr., Tiki Barber, Owen Wilson, and Justin Timberlake share their own "going green" stories about making a difference in the environment.

Steffen, A. *Worldchanging: A User's Guide for the 21st Century.* New York: Harry N. Abrams, 2006, 608 pp. ISBN-13: 978-0810930957

This book, a companion to the Web site www.worldchanging .com, is loaded with practical information on how to improve the environmental sustainability of our everyday lives and work toward a healthy, prosperous future. The book focuses on seven different topic areas: stuff, shelter, cities, community, business, politics, and the planet, describing ways to green each one.

Articles and Reports

Climate Science

Arctic Climate Impact Assessment: Impacts of a Warming Arctic. Cambridge, UK: Cambridge University Press, 2004, 144 pp. ISBN-13: 978-0521617789. Available online at http://amap.no/ acia.

This important government report summarizes the impacts that global climate change may have on the Arctic, which is extremely vulnerable to global warming. The report details 10 key findings on the climate change impacts in the region.

Hansen, J. E. "A Slippery Slope: How Much Global Warming Constitutes 'Dangerous Anthropogenic Interference'?" *Climatic Change* 68 (3): 269–279.

National Aeronautics and Space Administration (NASA) scientist James Hansen performed his own investigation comparing the levels of climate change presently occurring and the impacts this climate change will have on the melting of the polar ice sheets with projections from a 2001 IPCC report. His research found that the models used to simulate sea-level rise in the IPCC report failed to incorporate several processes that accelerate ice sheet disintegration over time, therefore underestimating by how much sea levels will rise.

Intergovernmental Panel on Climate Change (IPCC). *Climate Change 2007: The Physical Science Basis. Contribution of Working Group I of the Intergovernmental Panel on Climate Change,* edited by S. Solomon, D. Qin, M. Manning, Z. Chen, M. Marquis, K. B. Averyt, M. Tignor, and H. L. Miller. Cambridge, UK: Cambridge University Press, 2007, 996 pp. Also available online at http://ipcc-wg1.ucar.edu/wg1/wg1-report.html.

This important report is part of the IPCC Fourth Assessment Report. It contains the most up-to-date, comprehensive, and authoritative summary of climate change science. Included in the report are sections on the human and natural drivers of climate change, direct observations of recent climate change, a historical perspective of climate change, methods to understanding and attributing climate change, and projects of future changes in climate. It also provides critical information on the physical sciences basis for climate change.

Intergovernmental Panel on Climate Change (IPCC). *Climate Change 2007: Impacts, Vulnerability and Adaptation. Contribution of Working Group II of the Intergovernmental Panel on Climate Change.* **Cambridge, UK: Cambridge University Press, 2007. "Summary for Policymakers" available online at http://www.ipcc.ch/pdf/assessment-report/ar4/wg2/ar4-wg2-spm.pdf**

This report is also part of the IPCC Fourth Assessment Report. It provides a comprehensive summary of the impacts of climate change on natural and human systems, their vulnerability, and the capacity of these systems to adapt to climate change. Included in the report are sections on the current knowledge of the observed and future impacts of climate change, knowledge about responding to climate change, and systematic observing and research needs. Those seeking information on adaptation strategies to climate change should reference this report.

Scientific Expert Group on Climate Change (SEG), United Nations Foundation. *Confronting Climate Change: Avoiding the Unmanageable and Managing the Unavoidable.* **Report prepared for the UN Commission on Sustainable Development. Washington, DC: United Nations Foundation, 2007.**

This report offers recommendations for climate change mitigation and adaptation strategies and lays out a road map for policies that will allow for the implementation of improved adaptation and mitigation strategies.

Climate Policy and Economics

Climate Group. "Carbon Down, Profits Up." Surrey, UK: The Climate Group, 3rd Edition 2006, 32 pp. Available online at http://theclimategroup.org/assets/resources/cdpu_newedition.pdf.

"Carbon Down, Profits Up" outlines the efforts of companies and governments that have successfully managed their carbon footprints and, in many cases, created economic savings. Energy efficiency, renewable energy, and waste management were the three most widely used strategies used to reduce emissions.

Food and Agriculture Organization. "Adaptation to Climate Change in Agriculture, Forestry and Fisheries: Perspective, Framework and Priorities." Rome: Food and Agriculture Organization of the United Nations, 2007, 32 pp. Available online at ftp://ftp.fao.org/docrep/fao/009/j9271e/j9271e.pdf

This report describes the impacts that climate change will have on agriculture, forestry, and fisheries and discusses a framework for climate change adaptation strategies in these areas. Adaptation strategies mentioned in the report include seasonal crop rotation, forest fire management, and the use of a variety of new species. Also discussed is how the Food and Agriculture Organization is helping to implement some of these adaptation strategies.

Hoffman, A. *Getting Ahead of the Curve: Business Strategies on Climate Change.* Washington, DC: Pew Center on Global Climate Change, 2006, 137 pp. Available online at http://www.pewclimate.org/docUploads/PEW%5FCorpStrategies%2Epdf.

This report is an overview of the development and implementation of climate change strategies by businesses. The report was designed to serve as a how-to manual for other companies looking to put into practice similar climate change strategies and features an eight-step, three-stage plan to help businesses manage climate change. It has value not only for businesses but also for investors interested in evaluating corporate climate risk and policy makers seeking to learn how companies are reducing greenhouse gas emissions.

Innovest Strategic Value Advisors. "Carbon Disclosure Project Report 2007, Global FT500." London: Carbon Disclosure Project, 2007, 79 pp.; "Carbon Disclosure Project Report 2007, USA S&P 500FT500." London: Carbon Disclosure Project, 2007, 79 pp.

The Carbon Disclosure Project publishes yearly results of a survey of the greenhouse gas emissions and climate change activities of the world's largest companies. All reports are available at http://www.cdproject.net. The reports listed here summarize responses to the fifth Carbon Disclosure Project questionnaire and also address the risks and opportunities climate change presents. In addition, they provide an update on carbon disclosure and emission trends, explain the financial implications of climate change, and discuss climate change development, all with respect to business and the private sector.

Institute for Global Environmental Strategies (IGES). *CDM in Charts, Version 5.0.* **Kanagawa, Japan: IGES, 2007. Available online at http://www.iges.or.jp/en/cdm/pdf/charts.pdf.**

This publication provides an easy-to-read guide to the Clean Development Mechanism (CDM) of the Kyoto Protocol, including definitions of terms, a description of the steps involved in registering a project, and other vital information for anyone wishing to learn more about the process or considering developing a CDM project.

Intergovernmental Panel on Climate Change (IPCC). *Climate Change 2007: Mitigation. Contribution of Working Group III of the Intergovernmental Panel on Climate Change.* **Cambridge, UK: Cambridge University Press, 2007. Available online at http://www.mnp.nl/ipcc/pages_media/AR4-chapters.html.**

This report is part of the IPCC Fourth Assessment Report. Its focus is mitigation strategies and costs and covers topics including greenhouse gas emissions trends, mitigation in the short and medium term (until 2030), mitigation in the long term (beyond 2030), mitigation policies and instruments, sustainable development, and gaps in knowledge. The report is an excellent reference for those seeking to learn more about climate change mitigation strategies.

Intergovernmental Panel on Climate Change (IPCC). *Climate Change 2007: Synthesis Report.* **Cambridge, UK: Cambridge University Press, 2007. Available online at http://www.ipcc.ch/ipccreports/ar4-syr.htm**

The *Synthesis Report* is an integration of the reports of the three IPCC Working Groups. It summarizes the findings of their separate assessments and is the most comprehensive and authoritative review to date. It is essential reading for anyone interested in understanding climate change, the current and long-term impacts and risks, what can be done to mitigate emissions, and what can be done to adapt to impacts that have already begun to occur.

McKinsey and Company. *Reducing U.S. Greenhouse Gas Emissions: How Much at What Cost?* New York: McKinsey and Company and the Conference Board, 2007. Available online at http://www.conference-board.org/publications/describe.cfm ?id=1384.

International consulting group McKinsey and Company worked with leading institutions and experts to gather evidence from U.S. companies, academics, researchers, and interest groups to estimate the long-term costs and emission reduction impact of various options for reducing greenhouse gas emissions. Their findings include the central conclusion that the United States could reduce its emissions using proven and high-potential emerging technologies by 7 to 28 percent below 2005 levels by 2030, depending on the level of investment and political commitment, at a cost of less than $50 per ton of carbon reduced.

Pacala, S., and R. Socolow. "Stabilization Wedges: Solving the Climate Problem for the Next 50 Years with Current Technologies." *Science* 305: 968–972, 2004.

This classic article introduced the concept of "wedges" into discussions of climate policy. A stabilization wedge represents a certain amount of greenhouse gas emissions that must be eliminated in order to reduce emissions to acceptable levels. By matching up wedges with different technology changes, Pacala and Socolow outline strategies to reduce carbon dioxide emissions significantly using existing technologies.

Socolow, R. "Stabilization Wedges: An Elaboration of the Concept." In *Avoiding Dangerous Climate Change*, edited by H. J. Schellnhuber, W. Cramer, N. Nakicenovic, T. Wigley, and G. Yohe. Cambridge, UK: Cambridge University Press, 2006, pp. 347–354.

The author expands upon the argument that significant reductions are possible with existing technologies. The paper discusses stabilization wedges using energy efficiency, carbon capture and storage, nuclear energy, and renewable energy. It also draws attention to the need to use these stabilization wedges in developing countries and to how they can be utilized in a global carbon policy.

Socolow, R. H., and S. H. Lam. "Good Enough Tools for Global Warming Policy-Making." *Philosophical Transactions of the Royal Society* 365 (1853): 897–934, 2007.

This paper provides the quantitative tools for analyzing strategies to reduce greenhouse gas emissions. The paper, which incorporates the use of mathematical models, addresses three main questions: At what rates can carbon dioxide emissions continue once an established stabilization target for carbon dioxide levels has been reached? What time span should mitigation policies allow for the reduction of emissions to the target level to occur? How much harder will mitigation efforts be in the future if we fail to act now?

United Nations Development Programme (UNDP). "Human Development Report. Fighting Climate Change: Human Solidarity in a Divided World." New York: UNDP, 2007.

This report casts climate change as the defining human development issue of the 21st century. It tells how increased exposure to droughts, floods, and storms are destroying opportunity and reinforcing inequality. It argues that a failure to respond to these challenges will stall and eventually reverse international efforts to reduce poverty.

World Business Council on Sustainable Development (WBCSD). "Policy Directions to 2050: A Business Contribution to the Dialogues on Long-Term Action." Geneva: WBCSD, 2007, 52 pp. Available online at http://www.wbcsd.org/DocRoot/Grf5x6SMcHlMsH801XdE/int_low_res.pdf.

This report lays out a framework for climate change policies that will allow for sustained economic growth while transitioning to a low carbon economy, focusing on four priorities that should guide policy development.

Academic Journals

Bulletin of the American Meteorological Society
Publisher: American Meteorological Society
ISSN 1520-0477

Bulletin of the American Meteorological Society publishes articles on current research and advances in the fields of meteorology and climatology. The journal also features editorial commentary and book reviews.

Climatic Change
Publisher: Springer Netherlands
ISSN 0165-0009 (print), 1573-1480 (online)

Climatic Change focuses on climatic variability and change, including causes, impacts, scale, and implications. Articles include interdisciplinary research on topics such as meteorology, chemistry, physics, geography, economics, and policy analysis.

Climate Dynamics
Publisher: Springer
ISSN 0930-7575 (print), 1432-0894 (online)

An international research journal that focuses on all aspects of the dynamics of the global climate system, *Climate Dynamics* analyzes and models the atmosphere, oceans, biomass, and land surface as components of the climate system.

Climate Policy
Publisher: James & James, Ltd.
ISSN 1469-3062

Climate Policy publishes articles that summarize and analyze climate policy, including research on issues related to the UNFCCC, the Kyoto Protocol, international climate negotiations, and national policy.

Earth Interactions
Publisher: American Meteorological Society
ISSN 1087-3562

Earth Interactions features papers that explore the interactions among the biological, physical, and human components of the Earth system. Specifically, these papers explore the relationships among the hydrosphere, lithosphere, biosphere, and atmosphere, and place these interactions into the context of global climate change.

Ecological Economics
Publisher: Elsevier
ISSN 0921-8009

Ecological Economics focuses on extending and integrating the study and management of ecology and economics. The need for interaction between both environmental and economic policies is critical for sustainable development in the future. Selected research topics covered by papers in the journal include renewable resource management and conservation, methods to improve efficient environmental policies, and the implications of thermodynamics for economics and ecology.

Energy Policy
Publisher: Elsevier
ISSN 0301-4215

Energy Policy is a journal that focuses on the political, economic, planning, environmental, and social aspects of energy supply, demand, and utilization. Topics of papers that appear in the journal include national energy pricing, energy policy, and the environmental impacts of fossil fuel use. Papers specific to climate change included in the journal cover greenhouse gas mitigation, the IPCC reports, renewable energy, and strengthening energy systems in developing countries.

Geophysical Research Letters
Publisher: American Geophysical Union
ISSN 0094-8276

Geophysical Research Letters publishes short articles on specific issues within the geophysical sciences, which include the fields of meteorology and climate science. The short length of the papers allows for a quicker approval process, facilitating faster communication of the latest scientific research advances.

Global and Planetary Change
Publisher: Elsevier
ISSN 921-8181

Global and Planetary Change offers a multidisciplinary view of the causes, processes, and limits of variability in planetary change. Articles focus on the record of change in Earth history and the analysis and prediction of recent and future changes with specific discussion of changes in the chemical composition of the oceans and atmosphere, climate change, sea-level variations, human geography, and many other topics.

Global Change, Peace, and Security
Publisher: Taylor and Francis Group
ISSN 1478-1158

Global Change, Peace, and Security addresses the political, economic, and cultural questions posed by the rapid globalization being experienced by the world. The journal aims to explore trends in international issues such as human rights violations, environmental degradation, and exponential population growth.

Global Environmental Change
Publisher: Elsevier
ISSN 0959-3780

Global Environmental Change is an interdisciplinary journal that features research and review articles on scientific, policy, economic, and development issues related to global environmental change, including climate change.

Global Environmental Politics
Publisher: MIT Press
ISSN 1526-3800

Global Environmental Politics is the premier journal focused on international environmental policy and regularly contains articles that touch on climate change.

International Journal of Global Environmental Issues
Publisher: InderScience Publishers
ISSN 1466-6650

International Journal of Global Environmental Issues features articles on human environment, biodiversity, and issues surrounding global warming and climate change, which include the modeling of their impacts and the political conflicts they cause. The journal seeks to facilitate communication between experts from the government, academia, and the policy arena.

Journal of Applied Meteorology and Climatology
Publisher: American Meteorological Society
ISSN 1558-8432

This journal focuses on applied research related to topics such as physical meteorology, hydrology, weather modification, satellite meteorology, air pollution meteorology, agricultural and forest meteorology, and applied meteorological numerical models. Also included in the journal are articles covering research on applied climatology related to seasonal climate forecast applications and verification, climate risk and vulnerability, development of climate monitoring tools, and climate as it relates to the environment and society.

Journal of Atmospheric Sciences
Publisher: American Meteorological Society
ISSN 1520-0469

Journal of Atmospheric Sciences features articles that cover basic research related to the physics, dynamics, and chemistry of the atmosphere of the Earth and other planets, with emphasis on the quantitative and deductive aspects of the subject.

Journal of Climate
Publisher: American Meteorological Society
ISSN 1520-0442

Journal of Climate publishes articles on climate research, including those concerned with large-scale atmospheric and oceanic variability, changes in the climate system (including those caused by human activities), and climate simulation and prediction.

Journal of Environment and Development
Publisher: Sage
ISSN 1070-4965 (J339)

Journal of Environment and Development publishes research that focuses on the connection between environmental and development issues at all levels of society and government.

Journal of Environmental Economics and Management
Publisher: Elsevier Science
ISSN 0095-0696 (print), 1096-0449 (online)

Journal of Environmental Economics and Management features papers concerned with the linkage between economic systems and natural resources systems. It is regarded as one of the top journals in natural resource and environmental economics.

Journal of Geophysical Research Atmospheres
Publisher: American Geophysical Union
ISSN 0747-7309

Journal of Geophysical Research Atmospheres covers research on the physics and chemistry of the atmosphere, as well on the interaction between the atmosphere and the biosphere, lithosphere, and hydrosphere.

Nature
Publisher: Nature Publishing
ISSN 0028-0836

Perhaps the preeminent journal in the field, *Nature* publishes research articles that span a wide range of scientific fields. In many fields of science, new advances in research are introduced when they are published in this journal.

Science
Publisher: American Association for the Advancement of Science
ISSN 0035-8075

Science is the leading journal of scientific research, global news, and commentary. While much of the journal focuses on scientific research, additional topics include science policy and the impacts of science and technology.

Periodicals

Carbon Business
Publisher: Jeremy Blow
ISSN 1754 503X

Published quarterly, *Carbon Business* was created to provide businesses with information on ways to reduce their exposure to climate change risks and take advantage of new opportunities created by climate change policy and carbon markets. The focus is on how companies can gain comparative advantage as climate change policy develops.

Carbon Finance
Publisher: Fulton
ISSN 1742-5816

Carbon Finance is a monthly newsletter and e-mail update service providing in-depth coverage of the global markets in greenhouse gas emissions. It is fairly technical, with specific information on developments in the carbon markets as well as policy developments that affect the market.

Environment: Science and Policy for Sustainable Development
Publisher: Heldref
ISSN 0013-9157

Published six times a year, *Environment* is aimed at a more general audience than most academic journals but offers peer-reviewed articles and commentaries from researchers and practitioners who provide a broad range of international perspectives on issues related to sustainable development. Articles about various aspects of climate change appear frequently.

Internet Resources

Carbon Dioxide Information Analysis Center
http://cdiac.esd.ornl.gov/

The primary climate change data and information analysis unit of the U.S. Department of Energy, the Carbon Dioxide Information Analysis Center has data on greenhouse gas emissions and atmospheric concentrations, long-term climate trends, and the effect of elevated CO_2 on vegetation, among other topics.

Center for International Earth Science Information Network
http://ciesin.columbia.edu/

Columbia University's Center for International Earth Science Information Network works at the intersection of the social, natural, and information sciences. The Web site provides access to a vast amount of scientific and policy information related to human interactions with the environment.

Ceres
http://www.ceres.org

Ceres (pronounced "series") is a national network of investors, environmental organizations and other public interest groups working with companies and investors to address sustainability challenges such as global climate change. The Web site contains a vast amount of information on the economic risks associated with climate change and how investors and companies can and are responding.

ChinaDialogue
http://www.chinadialogue.net/tag/Climate_change

ChinaDialogue is an independent Web site maintained by a London-based educational charity. It provides bilingual information on China's environmental problems, including climate change.

Climate Action Network
http://www.climatenetwork.org/

The Climate Action Network is a worldwide network of more than 430 nongovernmental organizations that works to promote government and individual action to limit human-induced climate change to ecologically sustainable levels. Its Web site has papers, policy information, and guides to participation.

Discovery of Global Warming
http://www.aip.org/history/climate/

This Web site is an expanded version of Spencer Weart's book of the same name. It tells the history of climate change research as a single connected narrative, detailing the various collaborative efforts that led to the state of scientific knowledge we have today.

EarthWire Climate
http://www.earthwire.org/climate/

Earthwire is hosted by a branch of the United Nations Environment Programme (UNEP) called GRID-Arendal, whose mission is to provide credible, understandable environmental information to the public. The Earthwire Web site posts daily digests of news articles about climate change and related issues from around the world. Users can also subscribe to receive daily e-mails containing this information.

Environmental Change Institute, University of Oxford
http://www.eci.ox.ac.uk/

This Web site provides an interactive way to understand more about the climate system, global warming, and climate change. To illustrate the difference between weather and climate, a simulation that allows the user to roll dice is used. Additional animations and illustrations make this well-organized site a fun resource.

Global Change Master Directory
http://gcmd.gsfc.nasa.gov/

This NASA Web site contains a comprehensive directory of Earth science and global change data.

Global Methodology for Mapping Human Impacts on the Biosphere (GLOBIO)
www.globio.info

GLOBIO provides information on the impact of human activity of the biosphere. The program is run by UNEP; the Web site has links to publications including "Our Precious Coasts: Marine Pollution, Climate Change and Resilience of Coastal Ecosystems" (edited by C. Nellemann and E. Corcoran. Arendal, Norway: UNEP/GRID-Arendal, 2006).

Global Warming Archive
http://www.globalwarmingarchive.com

This Web site archives newspaper articles about global warming from around the world, providing a comprehensive look at public narratives about changes in climate from the 18th century through today.

Goddard Institute for Space Studies (GISS)
www.giss.nasa.gov

NASA's GISS conducts research on space and Earth science, addressing changes in our environment that affect the habitability of the planet. The Web site includes research, data, images, publications, and various software tools.

Greenhouse Gas Protocol Initiative
http://www.ghgprotocol.org/templates/GHG5/layout.asp?
type=p&MenuId=ODg4

The most widely used set of standards for inventorying greenhouse gas emissions, the Greenhouse Gas Protocol provides standards and guidance for companies and others developing a greenhouse gas inventory in addition to calculation tools.

Intergovernmental Panel on Climate Change
http://www.ipcc.ch/

The Intergovernmental Panel on Climate Change is a scientific body tasked with evaluating the risks of climate change caused by human activity. Its Web site maintains a wide number of publications produced by this Nobel Prize–winning organization.

International Petroleum Industry Environmental Conservation Association (IPIECA)
http://www.ipieca.org/index.php

IPIECA is the only global association representing both the upstream and downstream oil and gas industries on key global environmental and social issues. Its Climate Change Working Group focuses both on scientific assessments by the IPCC and the negotiations under the UNFCCC. It provides information and analysis to members and interested parties.

International Research Institute for Climate and Society (IRI)
http://portal.iri.columbia.edu/portal/server.pt

The mission of the IRI is to enhance society's ability to understand, anticipate, and manage the impacts of climate fluctuations and thereby improve human welfare and the health of the environment. Its Web site maintains an extensive library of climate-related data pertaining to both El Niño and long-term climate change, as well as information on the impacts, especially in developing countries, and efforts to develop effective adaptation strategies.

Lamont-Doherty Earth Observatory of the Earth Institute at Columbia University
http://www.ldeo.columbia.edu/

The Lamont-Doherty Earth Observatory is a leading research institution where research scientists seek fundamental knowledge about the origin, evolution, and future of the natural world. The Web site maintains a list of publications as well as a number of software tools to aid data analysis.

Massachusetts Institute of Technology (MIT) Joint Program on the Science and Policy of Global Change
http://globalchange.mit.edu

The MIT Joint Program on the Science and Policy of Global Change is an interdisciplinary organization that conducts research and policy analysis on issues relating to global change. The Web site maintains a large number of publications as well as modeling tools.

NASA Earth Observatory
http://earthobservatory.nasa.gov/Library/GlobalWarming Update/

This Web site features articles on global warming and climate change, with explanations of their causes, evidence they are occurring, how to model them, and the future impacts they may bring. Each section of the Web site has graphs, photographs, and charts to help explain the text on the page.

National Center for Atmospheric Research and the University Corporation of Atmospheric Research
http://www.eo.ucar.edu/basics/index.html

This Web site provides an overview of the basics of weather and climate change. Most sections of the Web site have photographs, charts, or graphs that help to reinforce the information being detailed. It also provides a list of links to other Internet resources on climate change.

National Oceanic and Atmospheric Administration (NOAA)
www.noaa.gov

The National Oceanic and Atmospheric Administration conducts research and gathers data on the world's oceans, atmosphere, space, and sun. A great deal of this research, along with data and images, is included on the Web site. NOAA also maintains the National Climatic Data Center, where a wide variety of climatic resources are available.

Pew Center for Climate Change
http://www.pewclimate.org/global-warming-basics/

This Web site contains information on the basics of global warming and climate change, including fact sheets available for download, sections that answer frequently asked questions and clarify myths, and a kids' page. The main part of the Pew Center Web site is home to numerous climate change reports available for the public to read online.

RealClimate
http://www.realclimate.org/

RealClimate is maintained by working climate scientists and geared toward the interested public and journalists. The site provides quick responses and context to developing climate stories.

Responding to Climate Change
http://www.rtcc.org/

Responding to Climate Change is an official observer to the United Nations Framework Convention on Climate Change process. Its Web site includes information on policies, innovations, technologies, products, and services related to the environmental issues and adaptation to the impacts of climate change.

See for Yourself
http://www.climate.yale.edu/seeforyourself/index.php

The interface allows users to choose assumptions about energy use and other factors. Based on those assumptions, the interface calculates economic impacts of reducing greenhouse gas emissions by certain percentages as predicted by a synthesis of 25 climate models being used to predict the economic impacts of reducing U.S. carbon emissions.

Tyndall Centre for Climate Change Research
http://www.tyndall.ac.uk/

The Tyndall Centre engages in cross-disciplinary research of sustainable responses to climate change. The Web site contains a number of publications and links to sister organizations.

United Nations Framework Convention on Climate Change (UNFCCC)
http://unfccc.int/2860.php

The Web site of the UNFCCC maintains important information on the evolution of international climate change talks. It also provides greenhouse gas emission data, national coping strategies, and plans for adaptation.

U.S. Climate Change Science Program
http://www.climatescience.gov/

The Climate Change Science Program integrates federal research on climate and global change. It incorporates information and data from the U.S. Global Change Data and Information System (www.globalchange.gov) and the U.S. Global Change Research Information Office (http://www.gcrio.org/); the Web site also has a great deal of information on U.S. climate policy.

U.S. Environmental Protection Agency (EPA) Climate Change Site
http://www.epa.gov/climatechange/

EPA's Climate Change Web site offers comprehensive information on the issue of climate change in a way that is accessible and meaningful to different segments of society, including communities, individuals, businesses, states and localities, and governments.

U.S. Geological Survey Earth Surface Dynamics Program
http://geochange.er.usgs.gov/

This program anticipates future environmental change by understanding the relationships between Earth surface processes, ecological systems, and human activities. It includes data sets and a great deal of information on the program's various projects.

World Resources Institute Climate Analysis Indicator Tool (CAIT)
http://cait.wri.org/

CAIT provides a comprehensive and comparable database of greenhouse gas emissions data (including all major sources and sinks) and other climate-relevant indicators. It can be used to analyze a wide range of climate-related data questions and to help support future policy decisions made under the Climate Convention and in other forums.

Film and Video

CBS News Presents: Global Warming (2007)
CBS News

This compilation contains 10 segments from *60 Minutes* and the *CBS Evening News* archive that chronicle the research, discoveries, and political implications of the Global Warming question.

Climate Change and Africa (2006)
http://www.youtube.com/watch?v=i9yMessHMXM

This video clip presents the problem faced by poor subsistent farmers of Africa due to global warming. Changes to climate as a result of global warming have caused rivers to dry, increasing degradation and widespread famine.

The Disappearing of Tuvalu: Trouble in Paradise (2004)
Directed by Christopher Horner and Gilliane Le Gallic
Distributed by Documentary Educational Resources
http://der.org/films/the-disappearing-of-tuvalu.html

This full-length documentary examines the tiny island nation of Tuvalu, which may become the first entire nation of environmental refugees because of climate change. Tuvalu is made up of a series of low-lying islands about 1,000 kilometers north of Fiji in the Pacific Ocean and is threatened by rising sea levels. In interviews with residents, politicians, activists, government workers, and scientists, the film documents how Tuvalu is struggling to cope with the effects of climate change. This film attempts to raise awareness about the country's problem to the rest of the world.

Earth System: Ice and Global Warming (2006)
WGBH Educational Foundation
http://www.teachersdomain.org/resources/ess05/sci/ess/eart hsys/esglaciers/index.html

This streaming Internet video describes the disproportionate impact that global warming has on ice- and snow-covered regions of the world as glaciers recede or fall into the ocean and sea ice melts. The Web site on which it is located also provides useful discussion questions and background reading to help students gain the most from the video.

Glacial Retreat in Greenland (2006)
Greenpeace International
http://www.greenpeace.org/australia/resources/videos/climat e-change/glacial-retreat-in-greenland

This brief Internet streaming video recounts with stunning visuals of landscapes the gradual but significant retreat of glaciated area in Greenland. The film attributes the event to global warming and includes the testimony of scientists to support the claim, but it ignores other important impacts or ways to resolve warming. It can be viewed for free on the Greenpeace Australia Pacific Web site.

Global Warming: The Signs and the Science **(2005)**
PBS
http://www.pbs.org/previews/globalwarming/

This film uses interviews with scientists to present evidence that global climate change is indeed taking place. This film does an excellent job of switching between a large-scale global perspective of climate change effects and a small-scale local perspective. Interviews with farmers, local school children, and others drive home the message that this problem affects the lives of everyday people.

Global Warming: What's Up with the Weather? **(2007)**
NOVA/Frontline
http://www.pbs.org/wgbh/warming/

NOVA and Frontline join forces on this two-hour special to explore the science and politics surrounding global warming. The film includes interviews with leading scientists, policy makers, and energy industry representatives with differing viewpoints on man's role and their predictions for the future.

A Global Warning? **(2007)**
A&E Home Video

First aired on the History Channel, this video provides a comprehensive look at the science of global warming and likely outcomes.

The Great Warming **(2006)**
Stonehaven Productions
http://www.thegreatwarming.com/

A documentary film about climate change narrated by Keanu Reeves and Alanis Morissette, *The Great Warming* covers the science, impacts, ethical and religious implications, and possible solutions to climate change.

"Hot Politics" (2007)
PBS
http://www.pbs.org/wgbh/pages/frontline/hotpolitics/view/

This episode of the PBS series *Frontline* explores the failure of the U.S. government to ratify the Kyoto Protocol or to take domestic measures to curb the emission of greenhouse gases. In interviews with policy makers, the show looks at the influence of the energy lobby, the manipulation of science by the government, and other aspects of the U.S. government's failure to act against what most of the world's governments see as an urgent global threat.

"Hot Times in Alaska" (2005)
Connecticut Public Television in association with *Scientific American* magazine
http://www.pbs.org/saf/1404/features/publicpolicy.htm

This episode of *Scientific American Frontiers* interviews scientists who are studying different species in Alaska and monitoring their responses to climate change. They talk about changes in trees, ice, vegetation, and invasive species. This episode is a comprehensive biological review looking at climate change and using case study examples and is a very entertaining and informative piece of media.

An Inconvenient Truth (2006)
http://www.climatecrisis.net/

Narrated by former U.S. vice president Al Gore, this documentary traces his concern with climate change from his early days in politics through the aftermath of the 2000 presidential election to Hurricane Katrina and beyond. It uses charts, graphs, and time-lapse photos of the Earth to provide evidence of global climate change. To combat global warming, the documentary suggests scientific, individual, and political action.

Inuit Observations of Climate Change (2008)
http://www.iisd.org/casl/projects/inuitobs.htm

In light of the dramatic changes that Inuit people have observed, the International Institute for Sustainable Development initiated a year-long project to document the problem of Arctic climate change. During this initiative, the project team produced a broadcast-quality video, *Inuit Observations on Climate Change,* which follows local people on land and at sea as they perform their traditional activities. Their voices—and the beauty of Canada's High Arctic—leave viewers with a new understanding of what will be lost if climate change continues unabated.

A King Tide Lashes Kiribati (2006)
Greenpeace International
http://www.greenpeace.org/australia/resources/videos/climate-change/rising-tide-threatens-pacific

This online streaming video shows the threat that global warming poses to island nations in the Pacific, focusing specifically on Kiribati. The film emphasizes the political struggle behind the issue, as representatives from small-island nations give frustrated interviews telling of developed nations' reluctance to acknowledge the gravity of global warming, their role in it, and their responsibility to help those who will be significantly impacted.

Melt: A Teenager's View of Global Warming (2006)
http://video.google.com/videoplay?docid=5617318652349295608

Ruby Reynolds, a 14-year-old filmmaker from the United Kingdom, explores how humans have contributed to global warming and what we should do to slow it.

Meltdown: A Global Warming Journey (2006)
http://ffh.films.com/id/12773/Meltdown_A_Global_Warming_Journey.html

By exploring contrasting points of view in the global warming debate, this film provides both political and scientific background to the climate change controversy.

Out of Balance: ExxonMobil's Impact on Climate Change (2006)
Joe Public Films
http://www.worldoutofbalance.org/

This documentary examines ExxonMobil's role in marketing fossil fuels as well as its active maneuvering to keep greenhouse gas emissions regulation off the political radar in the United States.

Patagonia's Glaciers Are Melting (2006)
Greenpeace International
http://www.greenpeace.org/australia/resources/videos/climat
e-change/patagonia-evidence-of-climat

This online streaming video gives a brief description with strik-
ing visual evidence of the receding glaciers in southern Chile.
The film includes interviews with area locals who have seen the
changes across generations, as well as scientists involved in the
generation of the IPCC report, all testifying to the reality and seriousness of global warming.

Planet in Peril (2007)
CNN
http://www.cnn.com/SPECIALS/2008/planet.in.peril/

This in-depth 3-hour documentary examines four key issues in
our changing planet: global warming, species loss, habitat loss,
and overpopulation. The program includes extensive footage
from remotes part of the Earth experiencing rapid change.

"Rewriting the Science" (2006)
CBS News
http://www.cbsnews.com/stories/2006/03/17/60minutes/mai
n1415985.shtml

In this video, which originally aired on the CBS news magazine
60 Minutes, the world's leading researcher on global warming
and head of NASA's GISS, Jim Hansen, explains how the Bush
administration has limited his freedom of speech. According to
Hansen, politicians have begun to rewrite climate science to the
detriment of the general public.

Six Degrees Could Change the World (2008)
National Geographic Video

Using stock footage and computer-generated images, this film provides viewers with a glimpse of the consequences of an increase in
global average temperatures of 6°C above the 20th-century average. Predictions vary widely, but unchecked greenhouse gas emissions could trigger climatic feedbacks that drive temperatures to
that point within a century. This film portrays the catastrophic effects that would likely accompany such an increase as well as the
tools at our disposal to avoid these outcomes.

Strange Days on Planet Earth: One Degree Factor (2005)
Sea Studios Foundation, Vulcan Productions, and National Geographic Television and Film
http://www.pbs.org/strangedays/episodes/onedegreefactor/
experts/frontlines.html

This documentary film demonstrates how changes to wildlife, the environment, and human health in very different parts of the world are all related to climate change. Hosted by Edward Norton, the film explores the effect that climate change will have on ocean food webs; the incidence of childhood asthma; Caribbean coral reefs; Lake Chad; and the Gwich'in tribe, a Native American group that hunts caribou for subsistence.

The 11th Hour (2007)
Warner Independent Pictures
http://wip.warnerbros.com/11thhour/mainsite/site.html

Narrated and produced by Leonardo DiCaprio, this film examines the ecological impacts of the modern, consumption-driven economy. Politicians, activists, and experts from a variety of fields provide insight into the limits of Earth's resources, what they see as the necessity for humans to adapt to those limits, and ideas and solutions that can help us move in that direction.

Too Hot Not to Handle (2006)
HBO Documentary Films
http://www.hbo.com/docs/programs/toohot/index.html

The HBO documentary *Too Hot Not to Handle* makes clear the overwhelming scientific consensus on two points: (1) Climate change is human induced and (2) climate change will have significant effects on the Earth's environment. The movie goes on to explore existing debates regarding how damaging these changes will be in the near and distant future.

Glossary

This chapter provides a brief introduction to key terms in climate science, policy, and economics. The first section lists some important climate-related acronyms. Those interested in an even more comprehensive list, and one that contains direct links to relevant home pages on the Internet, can consult the Climate Acronym Glossary maintained by the NASA Goddard Space Flight Center and currently available on the Web at http://climate.gsfc.nasa.gov/glossary.php. The second section contains definitions for important terms. Larger and more specialized lists are maintained online by the IPCC, EPA, Pew Center, and other organizations.

Climate-Related Acronyms

ACORE   American Council on Renewable Energy

AMAP   Arctic Monitoring and Assessment Programme

ANWR   Arctic National Wildlife Refuge

AOSIS   Alliance of Small Island States

Btu   British thermal unit

CAFE   corporate average fuel economy

CAN   Climate Action Network

CBD   Convention on Biological Diversity

CCS   carbon capture and storage/carbon capture and sequestration

CCX   Chicago Climate Exchange

CDM   Clean Development Mechanism

CDP   Carbon Disclosure Project

CER   certified emission reduction

CFCs   chlorofluorocarbons

CH_4 methane

CITES   Convention on International Trade in Endangered Species of Wild Fauna and Flora

CLRTAP   Convention on Long-Range Transboundary Air Pollution

CO_2 carbon dioxide

CO_2e carbon dioxide equivalent

COP   Conference of the Parties

CSD   Commission on Sustainable Development (United Nations)

DOE   U.S. Department of Energy

EC   European Community

EEA   European Environment Agency

ENSO   El Niño–Southern Oscillation

EPA   U.S. Environmental Protection Agency

ERU   emission reduction unit

EU   European Union

EU-ETS   European Union (greenhouse gas) Emissions Trading Scheme

FAO   Food and Agriculture Organization of the United Nations

FOE   Friends of the Earth

FTA   financial and technical assistance

G77   Group of 77 (developing-country negotiating bloc)

G8   Group of Eight (international forum for governments of the world's eight largest economies)

GATT   General Agreement on Tariffs and Trade

GCM   general circulation model

GDP   gross domestic product

GEF   Global Environment Facility

GHG   greenhouse gas

GISS   NASA Goddard Institute for Space Studies

GNP   gross national product

GROCC   Global Roundtable on Climate Change

GT   gigaton (1 billion metric tons)

HCFCs   hydrochlorofluorocarbons

HFCs   hydrofluorocarbons

IEA   International Energy Agency

IFF   Intergovernmental Forum on Forests

IGO   intergovernmental organization

IMF   International Monetary Fund

INC   Intergovernmental Negotiating Committee

IO   international organization

IPCC   Intergovernmental Panel on Climate Change

IRI   International Research Institute for Climate and Society

ITTA   International Tropical Timber Agreement

IUCN   World Conservation Union

JI   Joint Implementation

kWh   kilowatt hour

LDEO   Lamont-Doherty Earth Observatory, Columbia University

LEED   Leadership in Energy and Environmental Design (green building rating system)

LPG   liquid petroleum gas

LULUCF   Land use, land-use change, and forestry

MDB   multilateral development bank

MDGs   Millennium Development Goals

MEA   multilateral environmental agreement

MOP   Meeting of the Parties

mpg   miles per gallon

N_2O nitrous oxide

NAFTA   North American Free Trade Agreement

NASA   National Aeronautics and Space Administration (U.S.)

NGO   nongovernmental organization

NOAA   National Oceanic Atmospheric Administration (U.S.)

NOx   nitrogen oxides

NRDC   Natural Resources Defense Council

NWF   National Wildlife Federation

ODA   official development assistance

ODS   ozone-depleting substance

OECD   Organisation for Economic Co-operation and Development

OPEC   Organization of the Petroleum Exporting Countries

ppb   parts per billion

ppm   parts per million

R&D   research and development

RD&D   research, development, and deployment

RFF   Resources for the Future

RGGI   Regional Greenhouse Gas Initiative

SOx   sulfur oxides

SRI   socially responsible investing

tCO_2e ton of carbon dioxide equivalent

UCS   Union of Concerned Scientists

UN   United Nations

UNCCD   United Nations Convention to Combat Desertification

UNCTAD   United Nations Conference on Trade and Development

UNDP   United Nations Development Programme

UNEP   United Nations Environment Programme

UNFCCC   United Nations Framework Convention on Climate Change

UNFF   United Nations Forum on Forests

UNFPA   United Nations Population Fund

USGBC   U.S. Green Building Council

WBCSD   World Business Council for Sustainable Development

WHO   World Health Organization

WMO   World Meteorological Organization

WRI   World Resources Institute

WSSD   World Summit on Sustainable Development (2002)

WTO   World Trade Organization

WWF   World Wildlife Fund/Worldwide Fund for Nature

ZEV   zero emissions vehicle

Climate Terms

adaptation   The adjustment of human or natural systems in response to actual or expected climatic change in order to moderate harm or exploit beneficial opportunities.

aerosol   Solid or liquid particles suspended in the atmosphere.

afforestation   The planting of new forests on lands that have not recently been forested.

albedo   The ratio of the sun's light reflected by the Earth's surface to light received by it.

alternative energy   Any source of energy (e.g., solar, wind, geothermal) that can replace or supplement traditional fossil-fuel sources (such as coal, oil, and natural gas). Many of these sources are renewable and can be used indefinitely.

Annex I countries   The 40 countries listed in Annex I of the UNFCCC that agreed to try to limit their greenhouse gas emissions. The list includes industrialized countries that were members of the OECD in 1992, plus countries with economies in transition (including the Russian Federation, the Baltic states, and several Central and Eastern European states).

anthropogenic   Made by people or resulting from human activities. In the context of climate change, the term usually refers to greenhouse gas emissions that result from human activities including land-use change and the burning of fossil fuels.

atmosphere   The gaseous layers surrounding the Earth. The Earth's atmosphere is primarily composed of nitrogen (79 percent), oxygen (20 percent), small quantities of argon, and several greenhouse gases, including carbon dioxide, ozone, and water vapor.

biomass   The amount of living matter in a given unit of environmental area. Because organic matter can be converted to fuel, biomass is also regarded and often referred to in context as a potential energy source.

biosphere The part of the Earth and its atmosphere capable of supporting life.

carbon capture and storage (CCS) The process of capturing CO_2 from point sources that produce it and storing it underground, thereby avoiding atmospheric accumulation. Though large-scale capture and long-term storage are relatively untried, the process is regarded as a critical future tool for mitigating global warming.

carbon cycle The exchange of carbon between the biosphere, geosphere, hydrosphere, and atmosphere.

carbon dioxide (CO_2) A colorless, odorless gas formed during respiration, combustion, and organic decomposition. CO_2 is a naturally occurring component of the atmosphere and a powerful greenhouse gas. Human activities, including land-use change and the burning of fossil fuels, have increased atmospheric CO_2 levels dramatically since preindustrial times.

carbon dioxide equivalent (CO_2e) A measurement of the amount of a greenhouse gas other than CO_2 that describes the amount of CO_2 that would trap the same amount of heat over a specified period of time.

carbon sequestration The uptake and storage of carbon.

chlorofluorocarbons (CFCs) Compounds consisting of carbon, hydrogen, chlorine, and fluorine, once widely used as aerosol propellants and refrigerants. CFCs deplete the ozone layer when they break down, releasing chlorine into the stratosphere and harming the ozone layer.

Clean Development Mechanism (CDM) A provision of the Kyoto Protocol that allows industrialized countries to invest in projects that reduce greenhouse gas emissions in developing countries to meet emission reduction targets as an alternative to conducting more expensive emission reductions in their own countries.

climate The general pattern of weather conditions for a region over a long period of time (usually defined as greater than 30 years).

climate change Change in long-term trends in the average climate, including change to average temperatures and precipitation. It can be caused by natural factors or human activity.

climate sensitivity The equilibrium change in global mean surface temperature following a doubling of atmospheric CO_2 (equivalent) concentrations.

climate system Consists of the atmosphere, oceans, ice sheets, living organisms, soil, sediment, and rock, all of which affect the movement of heat around the Earth's surface.

Conference of the Parties (COP)   The decision-making body of the UNFCCC, made up of those governments (192 as of August 2008) that have ratified the convention. Other treaties and conventions also have conferences of the parties.

cryosphere   The portions of the Earth's surface where water is in solid form. This includes sea ice, lake ice, river ice, snow cover, glaciers, ice caps, ice sheets, and frozen ground.

deforestation   The conversion of forested areas to nonforested areas.

desertification   The progressive destruction or degradation of existing vegetative cover to form desert.

ecosystem   A community of organisms and its physical environment.

El Niño–Southern Oscillation (ENSO)   Sustained sea surface temperature anomalies of greater than 0.5°C across the central tropical Pacific Ocean; these anomalies have important consequences for weather around the globe.

emissions   The release of substances (e.g., greenhouse gases) into the atmosphere.

emissions cap   A mandated limit on the total amount of anthropogenic substances (e.g., greenhouse gases) that can be released into the atmosphere.

emissions trading   A market mechanism that allows emitters (countries, companies, or facilities) to buy or sell emissions from or to other emitters as part of a scheme to reduce overall emissions to an assigned level.

enhanced greenhouse effect   An increase in the natural greenhouse effect (see below) resulting from rising atmospheric greenhouse gas concentrations due to emissions from human activities.

entry into force   The point when an international agreement becomes legally binding for the governments that have ratified it. Terms for entry into force are written into individual treaties and vary considerably but usually are triggered when a prespecified number of parties have ratified the agreement.

environment   The air, water, minerals, organisms, and all other external factors surrounding and affecting a given organism at any time.

feedback (mechanism)   A process wherein the output of a system or process is looped back into the system or process itself.

forcing (mechanism)   A change in the balance of incoming radiation energy and outgoing radiation energy in a given climate system in response to an external influence. Frequently referred to as radiative forcing or climate forcing.

fossil fuels   Coal, oil, natural gas, and other hydrocarbons formed from the fossilized remains of dead plants and animals by exposure to heat and pressure in the Earth's crust over hundreds of millions of years. Burning fossil fuels can produce a number of detrimental environmental effects, including the production of greenhouse gases.

general circulation model (GCM)   A computer model designed to numerically simulate changes in climate as a result of slow changes in boundary conditions (such as the solar constant) or physical parameters (such as the greenhouse gas concentrations).

global warming   The gradual rise in the Earth's average surface temperatures, partly caused by human activity including the burning of fossil fuels.

global warming potential (GWP)   The cumulative warming effect exerted by a particular substance in the atmosphere, measured per unit over a specified period of time. The warming effect of a unit of emission of CO_2 is defined as 1.

greenhouse effect   The process by which atmospheric greenhouse gases trap heat flowing from the Earth's surface, preventing it from flowing into space. The natural greenhouse effect keeps the Earth's average temperature roughly 60°F warmer than it otherwise would be.

greenhouse gas (GHG)   Any atmospheric gas that can trap heat, warming the Earth's surface. Greenhouse gases include water vapor, carbon dioxide, methane, nitrous oxide, ozone, chlorofluorocarbons, and hydrofluorocarbons, among others.

gross domestic product (GDP)   The market value of all final goods and services produced within a country in a given period of time.

hydrofluorocarbons (HFCs)   Chemicals produced as substitutes for CFCs, particularly in refrigeration and semiconductor manufacturing. HFCs do not deplete the ozone layer; they are potent greenhouse gases (their Global Warming Potentials range from 1,300 to 11,700) and fall under the rules of the Kyoto Protocol.

hydrosphere   The component of the climate system composed of liquid surface and subterranean water, including oceans, seas, rivers, freshwater lakes, underground water, etc.

ice age   Any period of time in which glaciers dominate the surface of the Earth. The last major ice age occurred between 70,000 and 10,000 years ago.

ice core   A sample taken from snow and ice accumulated over a period of many years. The composition of ice cores—particularly the presence of hydrogen and oxygen isotopes—provides a picture of the climate at the time.

international organization (IO)/intergovernmental organization (IGO)
Any organization with sovereign states as members, such as the United
Nations.

Joint Implementation (JI) A provision of the Kyoto Protocol allowing
industrialized countries to invest in emissions-reducing projects in other
developed countries as an alternative to emission reductions in their
own countries.

Kyoto Protocol A treaty created under the UNFCCC that commits industrialized countries to legally binding targets to reduce their collective
greenhouse gas emissions at least 5 percent below 1990 levels by 2008 to
2012.

Little Ice Age A period of cooling that occurred between the 16th century and the mid-19th century, immediately following the medieval
warm period, most likely caused by a decrease in solar activity and increase in volcanic activity.

Medieval Warm Period A period of unusually warm climate in the
North Atlantic region, lasting from about the 10th century to about the
14th century.

methane (CH_4) A hydrocarbon greenhouse gas produced primarily by
respiration of anaerobic bacteria in places such as swamps, rice paddies,
and the intestines of ruminant animals; has a global warming potential
23 times that of carbon dioxide.

mitigation Taking actions that are aimed at reducing the extent of
global climate change, usually by reducing emissions or enhancing sinks.

Montreal Protocol The 1987 treaty that requires countries to eliminate
or severely restrict the production and use of all chemicals known to
damage the ozone layer, including CFCs.

negative feedback A process wherein the output of a system is looped
back into the system as an input and causes a reduction in the rate at
which the output is produced.

nitrous oxide (N_2O) A potent greenhouse gas produced primarily in
agriculture, particularly by the livestock sector; has a global warming
potential 296 times that of carbon dioxide.

nongovernmental organization (NGO) An organization whose members are private individuals, companies, or groups.

ozone (O_3) A reactive chemical compound consisting of three oxygen
atoms. Ozone in the stratosphere, known as the ozone layer, absorbs
incoming ultraviolet radiation from the sun, shielding the Earth's surface from this harmful radiation. Ozone in the troposphere is a greenhouse gas and a harmful air pollutant.

ozone depletion   Reduction in the concentration of stratospheric ozone caused by human use of chlorofluorocarbons and other chemicals that react with and destroy ozone. Ozone depletion is responsible for a seasonal hole in the ozone layer over Antarctica.

positive feedback   A process wherein the output of a system is looped back to become an input to the system and causes an increase in the rate at which the output is produced by the system.

radiation   Energy transfer in the form of electromagnetic waves or particles that release energy when absorbed by an object.

radiation balance   The balance between energy entering and leaving the Earth's atmosphere. Physics and thermodynamics dictate that, on average, equal amounts of energy enter and leave any isolated system, including the Earth and its atmosphere.

ratification   A state's official acceptance of an international agreement, which indicates its willingness to be bound by its provisions. A treaty comes into force only when a specified number of states have ratified it.

recycling   Collecting and reprocessing a resource that would otherwise be considered waste so that it can be used again.

reforestation   The replanting of forests on lands that have previously contained forests but have been converted to some other use.

(greenhouse gas) sink   Any process, activity, or mechanism that removes greenhouse gases, aerosols, or the precursors of greenhouse gases from the atmosphere.

stratosphere   The layer of the Earth's atmosphere that extends 12 kilometers to 40 kilometers above the surface of the Earth.

sustainable development   Balancing the fulfillment of human needs with the protection of the natural environment so that these needs can be met not only in the present but also in the indefinite future.

thermohaline circulation (THC)   The large-scale, density-driven circulation in the ocean caused by differences in temperature and salinity; an important component of the ocean-atmosphere climate system.

troposphere   The lowest layer of the Earth's atmosphere, extending from the Earth's surface to approximately 10 kilometers above it. The troposphere is where the clouds are found and the site of most weather phenomena.

ultraviolet (UV) radiation   Harmful electromagnetic radiation emitted from the sun with a wavelength between that of visible light and x-rays. Most UV radiation is blocked by ozone in the stratosphere.

United Nations Framework Convention on Climate Change (UN-FCCC)   A treaty signed at the 1992 Earth Summit that calls for the "stabilization of greenhouse gas concentrations in the atmosphere at a level that would prevent dangerous anthropogenic interference with the climate system."

urban heat island   The tendency for urban areas to have warmer air temperatures than the surrounding rural landscape due to the higher absorption of solar energy by buildings and asphalt.

vector-borne disease   Any disease that results from an infection transmitted to humans and other animals by blood-feeding arthropods, such as mosquitoes and ticks. Climate change may alter the frequency and distribution of these diseases.

water vapor   Water present in the atmosphere in gaseous form. The most abundant greenhouse gas and the one most responsible for the greenhouse effect, water vapor plays an important role in regulating the Earth's temperature.

weather   Describes the short-term (i.e., hourly and daily) state of the atmosphere at any given place. Weather is measured in variables including temperature, wind, cloudiness, and precipitation.

Index

About the Authors

Kate Brash is assistant director of the Global Roundtable on Climate Change, a project of the Earth Institute at Columbia University. She holds a graduate degree in environmental science and policy (Columbia University, 2004). She has previously worked as a writer and an editor and for nonprofit organizations.

David L. Downie is director of the Program on the Environment and associate professor of political science at Fairfield University. Dr. Downie's research focuses on the creation, content, and implementation of international environmental policy. The author of numerous publications on a variety of topics, his most recent works include *Global Environmental Politics, Fourth Edition* (Westview Press, 2006), with Pam Chasek and Janet Welsh Brown; *The Global Environment: Institutions, Law and Policy*, edited with Norm Vig and Regina Axelrod (CQ Press, 2004); and *Northern Lights against POPs: Combating Toxic Threats in the Arctic*, coedited with Terry Fenge (McGill-Queens University Press, 2003). Prior to joining Fairfield University in 2008, Dr. Downie taught courses in environmental politics at Columbia University from 1994 to 2008. While at Columbia, he also served as director of Environmental Policy Studies at the School of International and Public Affairs (1994–2000), as director of the Earth Institute Fellows Program (2002–2004), as associate director of the Graduate Program in Climate and Society (2004–2008), and as director of the Global Roundtable on Climate Change (2004–2008), among other positions.

Catherine Vaughan is a project coordinator at the International Research Institute for Climate and Society. She holds master's degrees in international relations (Yale University, 2007) and climate and society (Columbia University, 2008). She served with the U.S. Peace Corps in Zambia and as an adviser to the mission of Dominica to the United Nations. Her bachelor's degree, with high honors, is from Swarthmore College.